T-type Calcium Channels in Basic and Clinical Science

Stephen W. Schaffer • Ming Li
Editors

T-type Calcium Channels in Basic and Clinical Science

Springer

Editors
Stephen W. Schaffer
University of South Alabama
School of Medicine
Mobile, AL
USA

Ming Li
Department of Physiology
Tulane University School of Medicine
New Orleans, LA
USA

ISBN 978-3-7091-1412-4 ISBN 978-3-7091-1413-1 (eBook)
DOI 10.1007/978-3-7091-1413-1
Springer Wien Heidelberg New York Dordrecht London

Library of Congress Control Number: 2014957411

Printed on acid-free paper

Springer is part of Springer Science+Business Media (www.springer.com)

Preface

It has been 30 years since the initial characterization of T-type channel current by Armstrong and Matteson (1985) and 15 years since the molecular cloning of the three subtypes of the T-type Ca^{2+} channel by Perez-Reyes et al., Cribbs et al., and Lee et al. (1998, 1999). In the last decade much has been learned regarding the biophysics, biochemistry, physiology, and pharmacology of these channels, with information being gleaned from a wide range of studies varying from structure–function relationships to therapeutic applications. In this book, recent advances in T-type Ca^{2+} channel research are reviewed, with a special emphasis on the pathological implications of the channels and the potential role of channel antagonists in clinical medicine.

Part 1 provides a relatively detailed introduction to the T-type Ca^{2+} channels, with a focus on the biophysical properties of the channels, their regulation by intercellular and intracellular signaling pathways, and current attempts to develop more selective antagonists of the channel. In Chap. 1, Drs. Senatore and Spafford review key properties of the channels and summarize their role in cellular processes, such as excitation–secretion coupling and cell cycle control. In Chap. 2, the regulation of the T-type Ca^{2+} channels by a host of modulators, including glucose, cytokines, hormones, neurohumoral agents, metals, ATP, transcription factors, and common signaling pathways is discussed by Drs. Li and Wu. In Chap. 3, a fairly comprehensive review of recent and past T-type Ca^{2+} channel antagonists and their potential clinical value has been written by Drs. Kawazu and Hashimoto.

Part 2 reviews new information gleaned on the physiological functions of the T-type Ca^{2+} channels and their role in various clinical conditions. In Chap. 4, Drs. Warnier and coworkers review the role of various T-type Ca^{2+} channel subtypes in the differentiation of neuroendocrine cells and their role in excitation–secretion coupling. The possibility that the T-type Ca^{2+} channels might play a role in neuroendocrine tumors is also discussed. Chapter 5 focuses on the role of T-type Ca^{2+} channels in the development of cardiac hypertrophy and heart failure. Drs. Schaffer and Jong discuss the controversy surrounding the regulation of cardiac hypertrophy by T-type Ca^{2+} channels and the possibility that $Ca_v3.1$ and $Ca_v3.2$ might exert opposite actions on the development of heart failure. The topic of Chap. 6 concerns one of the initially described "physiological functions" of the T-type Ca^{2+} channels, namely, their contribution to diastolic depolarization. Dr. Schaffer also discusses recent data implicating T-type Ca^{2+} channels in the toxicity of Ca^{2+} overload related to electrical remodeling and ischemia–reperfusion injury. In Chap. 7, Drs. Chen and Parker review the convincing evidence that T-type Ca^{2+} channels play a central role in the development of absence seizures. It is anticipated that T-type Ca^{2+} channel antagonists will play a central role in the treatment of this condition. In Chap. 8, Drs. Pottle and Gray review the role of T-type Ca^{2+} channels in the regulation of cell cycling and proliferation. They introduce the novel concept that T-type Ca^{2+} channel antagonists might halt cell cycling near the G1/S checkpoint, rendering the cells more susceptible to S phase-specific chemotherapy. At the present time, this idea is being advanced to a phase Ib study in patients with recurrent high-grade glioma, a trial conducted in conjunction with the National Cancer Institute. In Chap. 9, Dr. Keyser discusses the mechanism by which T-type Ca^{2+} channels influence pain. The possibility that T-type Ca^{2+} channel antagonists might diminish both the sensation of pain and the progression of diseases, such as diabetic neuropathy, is discussed. In Chap. 10, the complex relationship between hyperglycemia, T-type Ca^{2+} channel activity, and insulin secretion is discussed. Dr. Li suggests that the T-type Ca^{2+} channels might provide a new target for the treatment of type 2 diabetes mellitus.

Skeptics maintain that potential adverse side effects of the T-type Ca^{2+} channel blockers will hamper wide acceptance of these therapeutic agents. They point out that the T-type Ca^{2+} channels are broadly distributed in a variety of human tissues, but in particular in vital organs, such as heart and thalamic neurons, where they exert important functions. However, at least three subtypes of the T-type Ca^{2+} channels are present in humans, where they serve different functions. Some of these subtype-specific actions are discussed in Part 2. It is likely that the development of subtype-specific antagonists will improve the efficacy while diminishing the toxicity of the newly developed antagonists.

The T-type Ca^{2+} channels, like all voltage-gated Ca^{2+} channels, conduct extracellular Ca^{2+} into the cytosol. Nevertheless, T-type Ca^{2+} channels, unlike the L-type Ca^{2+} channels, only carry a "transient and tiny" current. Yet, irregular expression of T-type Ca^{2+} channels has been implicated in a number of disease conditions, including cardiac hypertrophy (Chap. 5), ischemia–reperfusion injury (Chap. 6), absence seizures (Chap. 7), tumor cell proliferation (Chap. 8), and hyperinsulinemia (Chap. 10). Thus, there is an adequate therapeutic index window for T-type Ca^{2+} channel blockers to suppress pathological conditions while retaining the physiological function of the channels in normal tissue. In concluding remarks, Drs. Senatore and Spafford in Chap. 1 state that "new T-type channel drugs on the horizon have potential for disease treatment. The newer and more specific drugs will also contribute to a clearer understanding of how T-type channels function in normal cellular physiology, as well as in pathological conditions." We feel that the future of the T-type Ca^{2+} channel is bright.

Mobile, AL Stephen W. Schaffer
New Orleans, LA Ming Li

Contents

List of Contributors

Yucai Chen Department of Pediatrics, University of Illinois at Chicago, Peoria, IL, USA

Florian Gackière Laboratoire de Physiologie Cellulaire Inserm U1003, Université des Sciences et Technologies de Lille 1, Villeneuve d'Ascq, France

Laboratory of Excellence, Ion Channels Science and Therapeutics, Université des Sciences et Technologies de Lille 1, Villeneuve d'Ascq, France

Lloyd S. Gray Tau Therapeutics, LLC, Charlottesville, VA, USA

Norio Hashimoto Pharmaceutical Division, Nissan Chemical America Corporation, Houston, TX, USA

Chian Ju Jong Department of Pharmacology, University of South Alabama College of Medicine, Mobile, AL, USA

Takeshi Kawazu Pharmaceutical Research Department, Biological Research Center, Nissan Chemical Industries Ltd, Saitama, Japan

Brian M. Keyser Cellular and Molecular Biology Branch, U.S. Army Medical Research Institute of Chemical Defense Research Division, Aberdeen Proving Ground, MD, USA

Ming Li Department of Physiology, Tulane University Health Sciences Center, New Orleans, LA, USA

Pascal Mariot Laboratoire de Physiologie Cellulaire, Inserm U1003, Université des Sciences et Technologies de Lille 1, Villeneuve d'Ascq, France

Laboratory of Excellence, Ion Channels Science and Therapeutics, Université des Sciences et Technologies de Lille 1, Villeneuve d'Ascq, France

W. Davis Parker Jr Department of Pediatrics, University of Illinois at Chicago, Peoria, IL, USA

Jonathan E. Pottle Department of Physiology, Tulane University Health Sciences Center, New Orleans, LA, USA

Morad Roudbaraki Laboratoire de Physiologie Cellulaire, Inserm U1003, Université des Sciences et Technologies de Lille 1, Villeneuve d'Ascq, France

Laboratory of Excellence, Ion Channels Science and Therapeutics, Université des Sciences et Technologies de Lille 1, Villeneuve d'Ascq, France

Stephen W. Schaffer Department of Pharmacology, University of South Alabama College of Medicine, Mobile, AL, USA

Adriano Senatore Neuroscience Institute, Georgia State University, Atlanta, GA, USA

J. David Spafford Department of Biology, University of Waterloo, Waterloo, ON, Canada

Marine Warnier Laboratoire de Physiologie Cellulaire, Inserm U1003, Université des Sciences et Technologies de Lille 1, Villeneuve d'Ascq, France

Laboratory of Excellence, Ion Channels Science and Therapeutics, Université des Sciences et Technologies de Lille 1, Villeneuve d'Ascq, France

Songwei Wu Department of Anesthesiology and Perioperative Medicine, Georgia Regents University, Augusta, GA, USA

Part 1

General Properties of T-Type Ca^{2+} Channels

Physiology and Pathology of Voltage-Gated T-Type Calcium Channels

1

Adriano Senatore and J. David Spafford

Abstract

T-type channels are low voltage-activated members of the calcium channel family that also includes the high voltage-activated Ca_v1 and Ca_v2 channels. T-type channels open with only minimal depolarization or in response to hyperpolarization of the cell membrane and are associated with regulating excitability and pacemaking at subthreshold voltages. Interestingly, increasing evidence suggests that the subthreshold properties of T-type channels are exploited for other cellular processes including low-threshold synaptic vesicle release (excitation-secretion coupling), myocyte contraction and tone (excitation-contraction coupling), and cell cycle control. T-type channels are implicated in several pathologies including epilepsy, autism, sleep disturbances, pain, hypertension, and cancer. With the advent of novel blockers selective for T-type channels, their important contributions to normal cellular/organismal physiology, as well as to pathology, are becoming clearer.

1.1 Structural Features of T-Type Channels

T-type channels are members of a family of inward-permeating cation channels that includes voltage-gated sodium (Na_v) and calcium (Ca_v) channels. These channel types are made up of a single polypeptide bearing four homologous domains, each with six transmembrane alpha-helices homologous to those of the single subunit of a tetrameric voltage-gated potassium channel (segments S1–S6; Fig. 1.1). Segments S1–S4 make up the voltage sensor module of each domain (Fig. 1.1),

A. Senatore
Neuroscience Institute, Georgia State University, Atlanta, GA 30302, USA

J.D. Spafford (✉)
Department of Biology, University of Waterloo, Waterloo, ON, Canada N2L 3G1
e-mail: spafford@uwaterloo.ca

S.W. Schaffer, M. Li (eds.), *T-type Calcium Channels in Basic and Clinical Science*,
DOI 10.1007/978-3-7091-1413-1_1

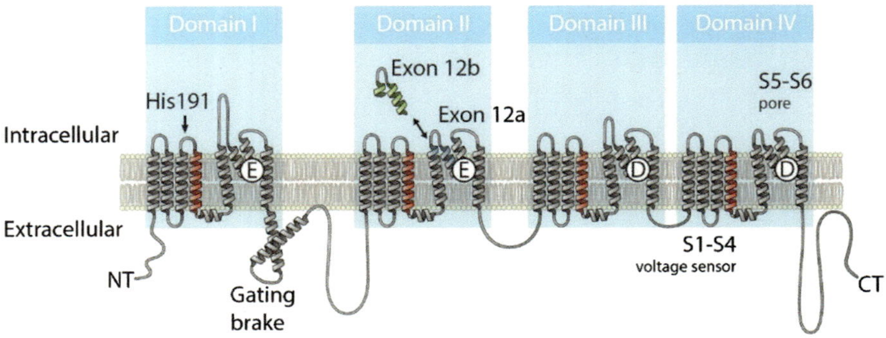

Fig. 1.1 Membrane topology of T-type calcium channels. T-type (or Cav3 type) calcium channels share the general structural features of four-domain voltage-gated channels including sodium channels (Na_v) and high voltage-activated calcium channels (i.e., Ca_v1 and Ca_v2 types). These channels are thought to have evolved via tandem duplication of a prokaryotic tetrameric voltage-gated channel gene, to produce a single large channel protein with four homologous domains (I–IV). Each domain contains six transmembrane segments/helices (S1–S6), with S1–S4 making up the voltage sensor module of each domain and S5 to S6 making up the pore module. S4 helices contain a series of positively charged lysine and arginine residues critical for voltage sensitivity. Distinguishing features for T-type channels include a selectivity filter of EEDD, contributed by four carboxyl-bearing residues (two glutamates and two aspartates) that project from the S5–S6 pore loops into the narrowest part of the pore to form a high-affinity binding site for Ca^{2+} (as compared to EEEE for Ca_v1 and Ca_v2 channels and DEKA for Na_v channels), and a gating brake structure on the cytosolic side of domain I S6, a helix-loop-helix motif that plays a critical role in T-type channel gating. The high-affinity binding site for block of $Ca_v3.2$ by Ni^{2+} and endogenous cation Zn^{2+} is largely attributed to an extracellular histidine residue located in the domain I S3–S4 loop (His191)

and together these endow channels with the ability to sense changes in voltage across the membrane to regulate opening/closing of the pore (Catterall 2010). The four pore modules consist of extracellular loops between transmembrane S5 and S6 helices that project one of four key amino acids into the pore to make up the "selectivity filter" motif (Fig. 1.1), a structure that largely governs what types of ions are allowed to pass through.

Ca_v channels exploit the high-affinity interaction between Ca^{2+} and negatively charged oxygen atoms in solution (Clapham 2007), by projecting an asymmetrical ring of carboxyl-bearing glutamate/aspartate residues into the pore, with selectivity filter motifs of "EEEE" for the HVA Ca_v1 and Ca_v2 channels and "EEDD" for T-type/Ca_v3 channels (Talavera and Nilius 2006). The resulting high-affinity binding site for calcium repels Na^+ ions from the pore at low concentrations of Ca^{2+}, rendering the channels calcium-selective (Yang et al. 1993; Cheng et al. 2010). In contrast, Na_v channels harbor a mix of negative (i.e., aspartate, glutamate), positive (lysine), and neutral (alanine) amino acids to make up a "DEKA" motif in vertebrates. Models of calcium and sodium ion selectivity have been derived from the X-ray coordinates of tetrameric channels, including K_v channels and prokaryotic Na_v channels (Doyle et al. 1998; Long et al. 2007; Payandeh et al. 2011), but these are symmetrical channels with selectivity filter residues that are positioned along a plane perpendicular to the permeation pathway, which is expected to be different relative to the asymmetrical 4-domain Ca_v and Na_v channels.

T-type channels appear to possess more ambiguous ion selectivity profiles when compared to the more highly selective pores of Ca_v and Na_v channels. Ca_v1 and Ca_v2 calcium channels pass almost exclusively calcium ions (>99.9 %) while excluding sodium ions, where this calcium can also serve as a cell messenger in converting electrical information (i.e., cellular excitability) into biochemical information (Clapham 2007) (i.e., structural changes in proteins, cell signaling) with roles such as in excitation-transcription, -secretion, and -contraction coupling. High levels of intracellular Ca^{2+} are toxic and can form precipitates in cells, so Na^+ as the main membrane depolarizing cation is better suited for purely electrogenic functions. Indeed, voltage-gated Na^+ channels (Na_v) that primarily shape the action potential spike bear highly sodium-selective pores that exclude Ca^{2+} ions almost completely (<0.1 %).

T-type channels have a more variable ion selectivity, where 20–25 % of permeating ions are Na^+ instead of Ca^{2+} for vertebrate $Ca_v3.1$ and $Ca_v3.2$ channels (Shcheglovitov et al. 2007). This Na^+ conductance is as high as 40 % of the total current for vertebrate $Ca_v3.3$ channels (Shcheglovitov et al. 2007). Invertebrate T-type channels have even greater sodium permeability, and in a subset of invertebrates (i.e., protostome such as mollusks, nematodes, and insects), alternative splicing in the domain II turret has evolved to alter Na^+ permeation from 50 % of the total current (mutually exclusive exon 12b; Fig. 1.1) to greater than 95 % (exon 12a Senatore et al. 2014). In the snail, the highly Na^+-permeant T-type isoforms expressed exclusively in the heart where they serve in lieu of absent Na_v channels for generating the main depolarizing Na^+ current for heart rhythmogenesis. The variability of the sodium permeability (20–95 %) in T-type channels suggests a flexibility in T-type channels in their capacity to serve the electrogenic roles of sodium channels or to contribute to Ca^{2+}-dependent intracellular processes.

T-type channels are also distinguished from other 4-domain calcium and sodium channels in not appearing to require accessory subunits for expression and function. Auxiliary beta subunits associate with a rigid alpha-helix that projects into the cytosol from the domain I, segment S6 transmembrane helix of Ca_v1 and Ca_v2 channels (termed that alpha-interaction domain or AID). T-type channels instead bear a helix-loop-helix structure in the analogous AID region, termed the "gating brake," which functions as an internal regulator for channel gating similar to how the β subunit regulates Ca_v1 and Ca_v2 channels (Perez-Reyes 2010). Mutations in the gating brake can cause excessive activity at hyperpolarized potentials (hence the "brake"), which in human $Ca_v3.2$ is genetically associated with childhood absence epilepsy Arias-Olguín et al. 2008).

1.2 T-Type Channels Are Optimized for Regulating Cellular Excitability

A major functional distinction between T-type channels and Ca_v1/Ca_v2 channels is their differing voltages of operation. Whereas Ca_v1 and Ca_v2 channels, in general, require much more potent depolarization to become activated, T-type channels can respond to only slight membrane potential changes. Thus, neurons and other cell types with abundant expression of T-type channels use them to regulate the overall excitability of the membrane, and T-type channels generally increase excitability. Paradoxically, however, T-type channels have a propensity to physically and/or functionally interact with their hyperpolarizing rivals, K^+ channels, leading to more complex contributions for T-type channels in regulating excitability. Mammalian $Ca_v3.2$ and $Ca_v3.3$ have been shown to directly couple to A-type (K_v4) potassium channels in cerebellar granule cells (Anderson et al. 2010), exerting a ~10 mV depolarizing shift in K_v4 steady-state inactivation. This interaction causes more A-type channels to be available for activation, effectively boosting membrane hyperpolarization and attenuating excitability. In substantia nigra dopaminergic (DA) neurons, T-type channels functionally associate with Ca^{2+}-sensitive small conductance K^+ channels (i.e., SK channels), where they increase SK channel inactivation to reduce K^+-conductance involved in action potential after-hyperpolarization, significantly altering the excitability and firing properties of the neurons (Wolfart and Roeper 2002). This apparent role for T-type channels as "regulators" of K^+ channels can also be reversed, as documented in prostate cancer cells where large conductance K^+ channels (i.e., BK channels) set the membrane potential to −40 mV, which is in the voltage range for constitutive calcium influx through the $Ca_v3.2$ T-type channel via what is termed a "window current" (Gackière et al. 2013).

Window currents are an interesting steady-state phenomenon for T-type channels, where within a small range of fixed voltages near the foot of the channel activation curve (Fig. 1.2), a small subset of the total T-type channel population is not inactivated and is thus able to conduct a small but constitutive inward current (Perez-Reyes 2003; Dreyfus et al. 2010; Senatore et al. 2012). Ca^{2+} window currents through T-type channels have been implicated in various cellular processes including cell division (and abnormal cell division in cancer) (Lory et al. 2006; Senatore et al. 2012; Gackière et al. 2013), as well as depolarizing the resting membrane potential (Dreyfus et al. 2010).

It is interesting that the same overlap between channel activation and inactivation that creates the window current also underlies another important function of T-type channels, post-inhibitory rebound (PIR) excitation. Here, membrane hyperpolarization via synaptic inhibition can drop the membrane potential enough to relieve inactivation of T-type channels, recruiting a large pool of available T-type channels to produce a robust post-inhibitory response (Perez-Reyes 2003). Thus, T-type channels can be activated by both depolarization, as noted above, but also hyperpolarization. One classical example of this occurs in the thalamus, where thalamocortical neurons depend on T-type channels for generating post-inhibitory

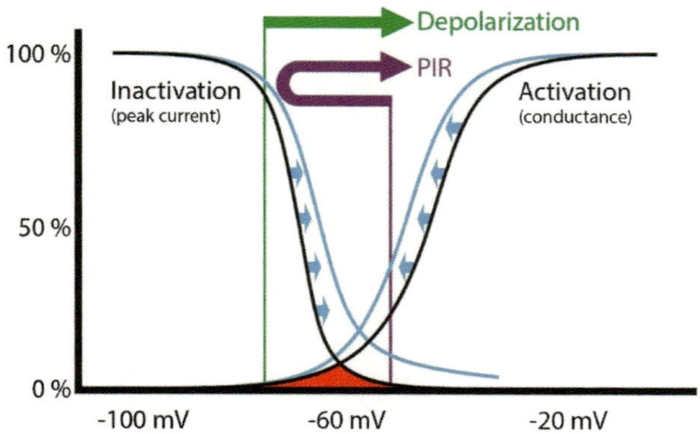

Fig. 1.2 Steady-state voltage properties of T-type channels. T-type channels operate in a subthreshold voltage range, with an overlap between steady-state activation and steady-state inactivation (*black curves*) producing a voltage range where a subset of T-type channels are constitutively active (i.e., the "window current," indicated in *red*). The overlap between T-type channels activation and inactivation also allows T-type channels to become activated not only by small depolarizations from rest (*green arrow*) but also by hyperpolarizing inputs that relieve inactivation allowing T-type channels to contribute to post-inhibitory rebound excitation (PIR; *purple arrow*). Alterations to the steady-state properties of T-type channels can have profound impacts on excitability. For example, knockout of PLCβ4 in thalamic neurons leads to a depolarizing shift in channel steady-state inactivation and a hyperpolarizing shift in channel activation (*blue curves* and *arrows*), allowing T-type channels to operate in a voltage range where they generally don't

"low-threshold calcium spikes" that project bursts of action potentials to the cortex to gate sensory information during non-REM sleep (Lee et al. 2004; Anderson et al. 2005; Crunelli et al. 2006). Similarly, Purkinje cell inhibitory input to deep cerebellar nuclear neurons (DCNs) recruits T-type channels for PIR excitation to produce high-frequency Na^+-dependent action potentials that form the main output of the cerebellum to other parts of the brain (Boehme et al. 2011).

1.3 Pharmacology of T-Type Channels

Our understanding of the functional roles for Ca_v1 and Ca_v2 channels have benefited greatly from the availability of potent and selective channel blockers (e.g., dihydropyridines for Ca_v1 channels and venom peptides such as ω-Agatoxin IVA and ω-Conotoxin for $Ca_v2.1$ and $Ca_v2.2$, respectively). Specific contributions of T-type channels have been difficult to assess because there has been a lack of specific blockers that will specifically block T-type channels or discriminate between the different vertebrate T-type channel subtypes (i.e., $Ca_v3.1$, $Ca_v3.2$, and $Ca_v3.3$). Ni^{2+} has classically been considered a T-type channel blocker, but only T-type currents generated by $Ca_v3.2$ channels are particularly sensitive (Kang et al. 2007). Mibefradil, used for treatment of angina and hypertension, was the only commercially available medicine marketed for its T-type channel blocking ability, but its negative side effects have precluded it as a viable medicine. Subsequently mibefradil, like other first generation T-type channel drugs, was found to not be selective for T-type channels and blocks other channel types such as sodium and other calcium channels. Newer T-type-specific blockers (e.g., TTA-P2/TTA-A2; NNC 55-0396; Z944; efonidipine) show greater blocking affinity for T-type channels yet still do not discriminate between mammalian subtypes ($Ca_v3.1$, $Ca_v3.2$, and $Ca_v3.3$).

1.4 Roles in Secretion and Contraction

A potential role for T-type channels in regulating secretion and exocytosis was suggested over 20 years ago in synaptic transmission between interneurons that regulate the leech heartbeat (Angstadt and Calabrese 1989). Since then, potential T-type channel involvement in mediating exocytosis has been implicated in several vertebrate tissues including retinal bipolar cells, granule cells of the olfactory bulb, hippocampal interneurons, nociceptive neurons, cortical neurons, and neuroendocrine cells (Carbone et al. 2006; Weiss et al. 2013; Weiss and Zamponi 2013). It has been hard to prove a direct role for T-type channels in secretion, because T-type channels can indirectly operate by depolarizing the membrane to activate high voltage-activated calcium channels that normally mediate secretion (Carbone et al. 2006; Escoffier et al. 2007). $Ca_v3.2$ channels in brain neurons and in neuroendocrine cells do bind to specific presynaptic SNARE proteins essential for vesicular release of transmitter, and this interaction is suggestive of a direct involvement of T-type channels in low-threshold exocytosis (Weiss et al. 2012). Similarly, a role for T-type channels in myocyte contraction and tone has long been postulated in both vertebrate and invertebrates (Yeoman et al. 1999; Ferron et al. 2002; Shtonda and Avery 2005; Ono and Iijima 2010), but a direct role has been debated (Cribbs 2006). The use of newer T-type channel blockers (TTA-P2; NNC 55-0396; efonidipine) has provided more concrete evidence that T-type channels participate in regulating pacemaker activity in contractile tissue such as vascular smooth muscle (Chen et al. 2003; Jensen and Holstein-Rathlou 2009; Cazade et al. 2013; Harraz et al. 2013), the uterus (Lee et al. 2009; Young et al. 1993), the bladder (Sui et al. 2009), and the urinary tract (Hurtado et al. 2014). T-type channels are also believed to play a prominent role in coupling membrane excitation to calcium-induced calcium release (CICR) for contraction in the mammalian embryonic heart, a configuration that is lost throughout development as cardiomyocytes develop transverse tubules that position L-type $Ca_v1.2$ channels in close proximity to Ca^{2+}-sensitive ryanodine receptors for CICR (Vassort et al. 2006; Ono and Iijima 2010).

1.5 Cell Division and Cancer

The window current of T-type channels provides a tonic depolarization and a steady-state flow of Ca^{2+} into the cytosol. Both the normal cell cycle and apoptosis have calcium-dependent checkpoints (Roderick and Cook 2008; Taylor et al. 2008b), and here T-type channels and specifically the trickle of calcium through window currents have been suggested as players in cell proliferation and growth under both normal conditions and cancer (Taylor et al. 2008a; Oguri et al. 2010; Li et al. 2011; Gackière et al. 2013). T-type channel expression spikes during G1 and S phases of the cell cycle in cultured smooth muscle cells (Taylor et al. 2008b), and a similar upregulation has been documented in various forms of cancer (Lory et al. 2006; Taylor et al. 2008b). For instance, the functional coupling of $Ca_v3.2$ T-type calcium channels and Ca^{2+}-sensitive BK K^+ channels appear to promote cellular proliferation in prostate cancer (Gackière et al. 2013). But notably, the contribution of T-type channels to cellular proliferation and various forms of cancer is inconsistent, and in some cases T-type channels appear to prevent proliferation (Toyota et al. 1999; Ueki et al. 2000; Shen et al. 2002; Diaz-Lezama et al. 2010). A working hypothesis is that the coupling of T-type channels to different cellular signaling pathways can be variable in different cell types, stages of development, and disease states, and these different configurations change qualitatively and quantitatively the contribution of T-type channels (Senatore et al. 2012).

1.6 Hypertension and Cardiac Hypertrophy

Recent findings indicated that T-type channels, at least in part, contribute to vascular tone. A recent study revealed that endogenous vasodilator nitric oxide (NO) attenuates T-type channel currents in rat cerebral arteries through a protein kinase G/guanylyl cyclase pathway, leading to a decrease in arterial myogenic tone (Harraz et al. 2013). A similar finding has been made in mice, where the endogenous epoxyeicosatrienoic acid 5,6-EET was found to selectively block $Ca_v3.2$ T-type calcium currents in mesenteric arteries to promote vasodilation (Cazade et al. 2013). T-type channels have also been implicated in cardiac hypertrophy of the mammalian heart. T-type channel expression drops dramatically during development in mammals (Ono and Iijima 2010) [and invertebrates (Senatore and Spafford 2012)], but both $Ca_v3.1$ and $Ca_v3.2$ become re-expressed during cardiac hypertrophy (Ono and Iijima 2010; Vassort et al. 2006; Cribbs 2010), where it seems that the two channel isotypes exert opposing effects: whereas $Ca_v3.2$ contributes to cardiac hypertrophy by activating the pro-hypertrophic calcineurin/NFAT pathway (Chiang et al. 2009), $Ca_v3.1$ antagonizes hypertrophy by opposing this same pathway (Nakayama et al. 2009).

1.7 Sleep Disturbances, Epilepsy, and Autism Spectrum Disorders

One of the most enriched regions for T-type channel expression is the thalamus (Talley et al. 1999; McRory et al. 2001), which plays a fundamental role in relaying/gating somatosensory information to the cortex and regulating the state of wakefulness, alertness, and sleep. Thalamic neurons can have multiple modes of excitability, and these modes are each intimately associated with the different states of consciousness. During non-REM sleep, thalamocortical neurons receive inhibitory hyperpolarizing inputs from GABAergic reticular thalamic nuclei neurons, de-inactivating T-type channels to produce post-inhibitory rebound excitation and a mode of excitability with characteristic low-threshold calcium spikes that gate sensory information from the cortex (Lee et al. 2004; Anderson et al. 2005; Crunelli et al. 2006). Knockout and targeted deletion of thalamic $Ca_v3.1$ in mice produces significant disturbances to non-REM sleep (Anderson et al. 2005). More recent studies, exploiting next-generation T-type blockers TTA-A2 and Z944, have called into question the role of T-type channels in sleep (Uebele et al. 2009; Kraus et al. 2010; Tringham et al. 2012), but it should be noted that these blockers are nonselective between the three mammalian Ca_v3 channel isotypes, which might confound results in vivo.

There is a comorbidity between sleep and epilepsy, and individuals stricken with various forms of epilepsy also tend to experience sleep pathologies (Manni and Terzaghi 2010). Conversely, sleep and sleep disturbances are known to trigger seizures (Parrino et al. 2012). The most likely convergence of sleep and epilepsy is in the thalamus where T-type channels facilitate sleep spindle waves and are the focal point of absence epileptic seizures that manifest as a "spike-wave discharge" of synchronous oscillatory activity measurable on electroencephalograms (EEGs). T-type channels are upregulated in the thalamus of rat and mouse models of absence epilepsy (Talley et al. 2000; Zhang et al. 2002; Broicher et al. 2008), which, given their role in boosting excitability, fits within a model where they contribute to the thalamic spike-wave discharge during absence epilepsy. Genetic screens have found associations between several missense polymorphisms of $Ca_v3.2$ and childhood absence epilepsy in humans (Liang et al. 2006, 2007); in vitro, these mutations significantly alter the channels' properties toward hyperexcitability (Vitko et al. 2005, 2007). A similar association study for human $Ca_v3.1$, however, failed to show any significant changes in channel properties (Singh et al. 2007). Knocking out $Ca_v3.1$ in mice results in resistance to drug-induced seizures, while overexpression of $Ca_v3.1$ can induce the recurrence of absence epileptic discharges (Ernst et al. 2009).

There is also a link between epilepsy and autism spectrum disorders (ASDs), heritable developmental disorders characterized by deficits in communication, social skills, and the prevalence of repetitive behaviors (Schmunk and Gargus 2013). The comorbidity is strong with 26 % of children with ASD aged 13 and older also having epilepsy (Viscidi et al. 2013). Unfortunately, few consistent genetic markers have been identified for ASD; however molecular determinants for aberrant calcium handling and neuronal hyperexcitability have emerged as commonalities (Schmunk and Gargus 2013). All three human Ca_v3 channel isoforms have been implicated in ASD which is consistent with their involvement in boosting excitability as occurs in epilepsy (Splawski et al. 2006; Strom et al. 2010; Lu et al. 2012).

1.8 Neuropathic Pain

$Ca_v3.1$ and $Ca_v3.2$ channels have a broad, overlapping expression in the central and peripheral nervous systems as well as various non-neuronal cell types including neuroendocrine cells, sperm, myocytes, and bone (Iftinca 2011). $Ca_v3.2$ channels are uniquely expressed in abundance in nociceptive dorsal root ganglion neurons (DRG) of the peripheral nervous system, where they play an important role in boosting synaptic transmission between DRG neurons and spinal cord interneurons that project to the thalamus (Zamponi et al. 2009). $Ca_v3.2$ channel gene transcription and membrane expression are upregulated in animal models of neuropathic pain and diabetic neuropathy (Jagodic et al. 2007, 2008; Wen et al. 2010). $Ca_v3.2$ knockout (Choi et al. 2007), as well as antisense knockdown (Bourinet et al. 2005; Marger et al. 2011) and pharmacological disruption (Marger et al. 2011; Matsunami et al. 2011) in the peripheral nervous system, all significantly attenuate pain consistent with a role for $Ca_v3.2$ in peripheral pain processing (Choi et al. 2007). Interestingly, the contribution of $Ca_v3.2$ to pain processing can be antagonistically modulated by endogenous cation Zn^{2+} and by redox modulation. Zn^{2+}, using the same high-affinity binding site for Ni^{2+} specific for $Ca_v3.2$ (histidine 191 in domain I S3-S4 linker (Kang et al. 2006, 2007); Fig. 1.1), effectively blocks $Ca_v3.2$ currents to attenuate pain, but introduction of endogenous reducing agent L-cysteine, which is expected to alter disulfide bonds on the extracellular surface of the channel, lowers Zn^{2+} binding affinity and effectively increases $Ca_v3.2$ channel activity and pain sensitivity (Nelson et al. 2007).

The role of T-type channels in pain processing appears to be reversed in the thalamus. T-type channels promote thalamic low-threshold spikes that shut down cortical somatosensory processing (including peripheral neuropathy) during non-REM sleep. During wakefulness, thalamocortical neurons are depolarized beyond the range of T-type channel activity, and they exhibit tonic firing toward the cortex with frequencies that reflect the intensity of incoming sensory inputs (Jones 2010). Thalamocortical neurons in turn receive feedback from glutamatergic cortical neurons, entraining them into the tonic firing mode (Jones 2010). Interestingly, disrupting corticothalamic metabotropic glutamate receptor signaling via knockout of phospholipase Cβ4 (PLCβ4) amplifies thalamocortical L-type and T-type channel currents and promotes low-threshold spikes during wakefulness, leading to attenuated pain sensitivity (Cheong et al. 2008). Here, the ability of T-type channels to contribute to excitability in a voltage range where they are normally inactivated is attributed to a depolarizing shift in their steady-state inactivation and a higher propensity to activate at negative potentials (Fig. 1.2) caused by reduced activity of PLCβ4 downstream effector protein kinase C, a modulator of T-type channel function (Cheong et al. 2008).

Conclusions

It was in 1975 that Susumu Hagiwara discovered the T-type channel current, distinguished as the leftward hump on a current-voltage curve in the voltage clamp of starfish eggs. More than two decades later, the first T-type channel genes were identified. Since then, progress has been slow in identifying the complete range of T-type channel roles because of an absence of selective blockers, a weak animal phenotype in single T-type channel gene knockout animals, and an inability to separate T-type channel currents from the many other subthreshold cation currents that are often found in the same cells. It has been increasingly clear that T-type channels are more than just contributors to pacemaker rhythms in the brain and heart, with other potential roles such as in low-threshold exocytosis, myocyte contraction and tone, and cell cycle control and development. New T-type channel drugs on the horizon have potential for disease treatment including epilepsy, autism, sleep disturbances, pain, hypertension, and cancer. The newer and more specific drugs will also contribute to a clearer understanding of how T-type channels function in normal cellular physiology, as well as in pathological conditions.

References

Anderson MP, Mochizuki T, Xie J, Fischler W, Manger JP, Talley EM, Scammell TE, Tonegawa S (2005) Thalamic Cav3.1 T-type Ca^{2+} channel plays a crucial role in stabilizing sleep. Proc Natl Acad Sci USA 102:1743–1748

Anderson D, Mehaffey WH, Iftinca M, Rehak R, Engbers JD, Hameed S, Zamponi GW, Turner RW (2010) Regulation of neuronal activity by Cav3-Kv4 channel signaling complexes. Nat Neurosci 13:333–337

Angstadt JD, Calabrese RL (1989) A hyperpolarization-activated inward current in heart interneurons on the medicinal leech. J Neurosci 9:2846–2857

Arias-Olguín II, Vitko I, Fortuna M, Baumgart JP, Sokolova S, Shumilin IA, Van Deusen A, Soriano-García M, Gomora JC, Perez-Reyes E (2008) Characterization of the gating brake in the I-II loop of $Ca_v3.2$ T-type Ca^{2+} channels. J Biol Chem 283:8136–8144

Boehme R, Uebele VN, Renger JJ, Pedroarena C (2011) Rebound excitation triggered by synaptic inhibition in cerebellar nuclear neurons is suppressed by selective T-type calcium channel block. J Neurophysiol 106:2653–2661

Bourinet E, Alloui A, Monteil A, Barrère C, Couette B, Poirot O, Pages A, McRory J, Snutch TP, Eschalier A, Nargeot J (2005) Silencing of the $Ca_v3.2$ T-type calcium channel gene in sensory neurons demonstrates its major role in nociception. EMBO J 24:315–324

Broicher T, Kanyshkova T, Meuth P, Pape H-C, Budde T (2008) Correlation of T-channel coding gene expression, I-T, and the low threshold Ca^{2+} spike in the thalamus of a rat model of absence epilepsy. Mol Cell Neurosci 39:384–399

Carbone E, Giancippoli A, Marcantoni A, Guido D, Carabelli V (2006) A new role for T-type channels in fast "low-threshold" exocytosis. Cell Calcium 40:147–154

Catterall WA (2010) Ion channel voltage sensors: structure, function, and pathophysiology. Neuron 67:915–928

Cazade M, Bidaud I, Hansen PB, Lory P, Chemin J (2013) 5,6-EET potently inhibits T-type calcium channels: implication in the regulation of the vascular tone. Pflügers Arch

Chen CC, Lamping KG, Nuno DW, Barresi R, Prouty SJ, Lavoie JL, Cribbs LL, England SK, Sigmund CD, Weiss RM, Williamson RA, Hill JA, Campbell KP (2003) Abnormal coronary function in mice deficient in α 1H T-type Ca^{2+} channels. Science 302:1416–1418

Cheng RC, Tikhonov DB, Zhorov BS (2010) Structural modeling of calcium binding in the selectivity filter of the L-type calcium channel. Eur Biophys J 39:839–853

Cheong E, Lee S, Choi BJ, Sun M, Lee CJ, Shin HS (2008) Tuning thalamic firing modes via simultaneous modulation of T- and L-Type Ca^{2+} channels controls pain sensory gating in the thalamus. J Neurosci 28:13331–13340

Chiang CS, Huang CH, Chieng H, Chang YT, Chang D, Chen JJ, Chen YC, Chen YH, Shin HS, Campbell KP, Chen CC (2009) The $Ca_V3.2$ T-type Ca^{2+} channel is required for pressure overload-induced cardiac hypertrophy in mice. Circ Res 104:522–530

Choi S, Na HS, Kim J, Lee J, Lee S, Kim D, Park J, Chen CC, Campbell KP, Shin HS (2007) Attenuated pain responses in mice lacking Ca(V)3.2 T-type channels. Genes Brain Behav 6:425–431

Clapham DE (2007) Calcium signaling. Cell 131:1047–1058

Cribbs LL (2006) T-type Ca^{2+} channels in vascular smooth muscle: multiple functions. Cell Calcium 40:221–230

Cribbs L (2010) T-type calcium channel expression and function in the diseased heart. Channels 4:447–452

Crunelli V, Cope DW, Hughes SW (2006) Thalamic T-type Ca^{2+} channels and NREM sleep. Cell Calcium 40:175–190

Diaz-Lezama N, Hernández-Elvira M, Sandoval A, Monroy A, Felix R, Monjaraz E (2010) Ghrelin inhibits proliferation and increases T-type Ca^{2+} channel expression in PC-3 human prostate carcinoma cells. Biochem Biophys Res Commun 403:24–9

Doyle DA, Morais Cabral J, Pfuetzner RA, Kuo A, Gulbis JM, Cohen SL, Chait BT, MacKinnon R (1998) The structure of the potassium channel: molecular basis of K + conduction and selectivity. Science 280:69–77

Dreyfus FM, Tscherter A, Errington AC, Renger JJ, Shin HS, Uebele VN, Crunelli V, Lambert RC, Leresche N (2010) Selective T-type calcium channel block in thalamic neurons reveals channel redundancy and physiological impact of I(T)window. J Neurosci 30:99–109

Ernst WL, Zhang Y, Yoo JW, Ernst SJ, Noebels JL (2009) Genetic enhancement of thalamocortical network activity by elevating α1g-mediated low-voltage-activated calcium current induces pure absence epilepsy. J Neurosci 29:1615–1625

Escoffier J, Boisseau S, Serres C, Chen CC, Kim D, Stamboulian S, Shin HS, Campbell KP, De Waard M, Arnoult C (2007) Expression, localization and functions in acrosome reaction and sperm motility of Ca(V)3.1 and Ca(V)3.2 channels in sperm cells: an evaluation from Ca(V)3.1 and Ca(V)3.2 deficient mice. J Cell Physiol 212:753–763

Ferron L, Capuano V, Deroubaix E, Coulombe A, Renaud JF (2002) Functional and molecular characterization of a T-type Ca^{2+} channel during fetal and postnatal rat heart development. J Mol Cell Cardiol 34:533–546

Gackière F, Warnier M, Katsogiannou M, Derouiche S, Delcourt P, Dewailly E, Slomianny C, Humez S, Prevarskaya N, Roudbaraki M, Mariot P (2013) Functional coupling between large-conductance potassium channels and Cav3.2 voltage-dependent calcium channels participates in prostate cancer cell growth. Biol Open 2:941–951

Harraz OF, Brett SE, Welsh DG (2013) Nitric oxide suppresses vascular voltage-gated T-type Ca^{2+} channels through cGMP/PKG signaling. Am J Physiol 306:H279–H285

Hurtado R, Bub G, Herzlinger DA (2014) Molecular signature of tissues with pacemaker activity in the heart and upper urinary tract involves coexpressed hyperpolarization-activated cation and T-type Ca^{2+} channels. FASEB J 28:730–739

Iftinca MC (2011) Neuronal T-type calcium channels: what's new? Iftinca: T-type channel regulation. J Med Life 4:126–138

Jagodic MM, Pathirathna S, Nelson MT, Mancuso S, Joksovic PM, Rosenberg ER, Bayliss DA, Jevtovic-Todorovic V, Todorovic SM (2007) Cell-specific alterations of T-type calcium current in painful diabetic neuropathy enhance excitability of sensory neurons. J Neurosci 27:3305–3316

Jagodic MM, Pathirathna S, Joksovic PM, Lee W, Nelson MT, Naik AK, Su P, Jevtovic-Todorovic V, Todorovic SM (2008) Upregulation of the T-type calcium current in small rat sensory neurons after chronic constrictive injury of the sciatic nerve. J Neurophysiol 99:3151–3156

Jensen LJ, Holstein-Rathlou NH (2009) Is there a role for T-type Ca^{2+} channels in regulation of vasomotor tone in mesenteric arterioles? Can J Physiol Pharmacol 87:8–20

Jones EG (2010) Thalamocortical dysrhythmia and chronic pain. Pain 150:4–5

Kang HW, Park JY, Jeong SW, Kim JA, Moon HJ, Perez-Reyes E, Lee JH (2006) A molecular determinant of nickel inhibition in Cav3.2 T-type calcium channels. J Biol Chem 281:4823–4830

Kang HW, Moon HJ, Joo SH, Lee JH (2007) Histidine residues in the IS3-IS4 loop are critical for nickel-sensitive inhibition of the Ca(v)3.2 calcium channel. FEBS Lett 581:5774–5780

Kraus RL, Li Y, Gregan Y, Gotter AL, Uebele VN, Fox SV, Doran SM, Barrow JC, Yang ZQ, Reger TS, Koblan KS, Renger JJ (2010) In vitro characterization of T-type calcium channel antagonist TTA-A2 and in vivo effects on arousal in mice. J Pharmacol Exp Ther 335:409–417

Lee J, Kim D, Shin H-S (2004) Lack of delta waves and sleep disturbances during non-rapid eye movement sleep in mice lacking α1G -subunit of T-type calcium channels. Proc Natl Acad Sci USA 101:18195–18199

Lee SE, Ahn DS, Lee YH (2009) Role of T-type Ca channels in the spontaneous phasic contraction of pregnant rat uterine smooth muscle. Korean J Physiol Pharmacol 13:241–249

Li W, Zhang SL, Wang N, Zhang BB, Li M (2011) Blockade of T-type Ca^{2+} channels inhibits human ovarian cancer cell proliferation. Cancer Investig 29:339–346

Liang J, Zhang Y, Wang J, Pan H, Wu H, Xu K, Liu X, Jiang Y, Shen Y, Wu X (2006) New variants in the CACNA1H gene identified in childhood absence epilepsy. Neurosci Lett 406:27–32

Liang J, Zhang Y, Chen Y, Wang J, Pan H, Wu H, Xu K, Liu X, Jiang Y, Shen Y, Wu X (2007) Common polymorphisms in the CACNA1H gene associated with childhood absence epilepsy in Chinese Han population. Ann Hum Genet 71:325–335

Long SB, Tao X, Campbell EB, MacKinnon R (2007) Atomic structure of a voltage-dependent K^+ channel in a lipid membrane-like environment. Nature 450:376–382

Lory P, Bidaud I, Chemin J (2006) T-type calcium channels in differentiation and proliferation. Cell Calcium 40:135–146

Lu AT, Dai X, Martinez-Agosto JA, Cantor RM (2012) Support for calcium channel gene defects in autism spectrum disorders. Mol Autism 3:18

Manni R, Terzaghi M (2010) Comorbidity between epilepsy and sleep disorders. Epilepsy Res 90:171–177

Marger F, Gelot A, Alloui A, Matricon J, Ferrer JF, Barrère C, Pizzoccaro A, Muller E, Nargeot J, Snutch TP, Eschalier A, Bourinet E, Ardid D (2011) T-type calcium channels contribute to colonic hypersensitivity in a rat model of irritable bowel syndrome. Proc Natl Acad Sci USA 108:11268–11273

Matsunami M, Kirishi S, Okui T, Kawabata A (2011) Chelating luminal zinc mimics hydrogen sulfide-evoked colonic pain in mice: possible involvement of T-type calcium channels. Neuroscience 181:257–264

McRory JE, Santi CM, Hamming KS, Mezeyova J, Sutton KG, Baillie DL, Stea A, Snutch TP (2001) Molecular and functional characterization of a family of rat brain T-type calcium channels. J Biol Chem 276:3999–4011

Nakayama H, Bodi I, Correll RN, Chen X, Lorenz J, Houser SR, Robbins J, Schwartz A, Molkentin JD (2009) Alpha1G-dependent T-type Ca^{2+} current antagonizes cardiac hypertrophy through a NOS3-dependent mechanism in mice. J Clin Invest 119:3787–3796

Nelson MT, Woo J, Kang HW, Vitko I, Barrett PQ, Perez-Reyes E, Lee JH, Shin HS, Todorovic SM (2007) Reducing agents sensitize C-type nociceptors by relieving high-affinity zinc inhibition of T-type calcium channels. J Neurosci 27:8250–8260

Oguri A, Tanaka T, Iida H, Meguro K, Takano H, Oonuma H, Nishimura S, Morita T, Yamasoba T, Nagai R, Nakajima T (2010) Involvement of CaV3.1 T-type calcium channels in cell proliferation in mouse preadipocytes. Am J Physiol 298:C1414–C1423

Ono K, Iijima T (2010) Cardiac T-type Ca^{2+} channels in the heart. J Mol Cell Cardiol 48:65–70

Parrino L, De Paolis F, Milioli G, Gioi G, Grassi A, Riccardi S, Colizzi E, Terzano MG (2012) Distinctive polysomnographic traits in nocturnal frontal lobe epilepsy. Epilepsia 53:1178–1184

Payandeh J, Scheuer T, Zheng N, Catterall WA (2011) The crystal structure of a voltage-gated sodium channel. Nature 475:353–358

Perez-Reyes E (2003) Molecular physiology of low-voltage-activated T-type calcium channels. Physiol Rev 83:117–161

Perez-Reyes E (2010) Characterization of the gating brake in the I-II loop of CaV3 T-type calcium channels. Channels 4:453–458

Roderick HL, Cook SJ (2008) Ca^{2+} signalling checkpoints in cancer: remodelling Ca^{2+} for cancer cell proliferation and survival. Nat Rev Cancer 8:361–375

Schmunk G, Gargus JJ (2013) Channelopathy pathogenesis in autism spectrum disorders. Front Genet 4:222

Senatore A, Spafford JD (2012) Gene transcription and splicing of T-type channels are evolutionarily-conserved strategies for regulating channel expression and gating. PLoS One 7:e37409

Senatore A, Zhorov BS, Spafford JD (2012) Ca_V3 T-type calcium channels. WIREs Membr Transp Signall 1:467–491. doi:10.1002/wmts.41

Senatore A, Guan W, Boone AN, Spafford JD (2014) Ca_V3 T-type calcium channels become sodium channels using alternate extracellular turret residues outside the selectivity filter. J Biol Chem 289(17):11952–11969

Shcheglovitov A, Kostyuk P, Shuba Y (2007) Selectivity signatures of three isoforms of recombinant T-type Ca^{2+} channels. Biochim Biophys Acta 1768:1406–1419

Shen L, Ahuja N, Shen Y, Habib NA, Toyota M, Rashid A, Issa JP (2002) DNA methylation and environmental exposures in human hepatocellular carcinoma. J Natl Cancer Inst 94:755–761

Shtonda B, Avery L (2005) CCA-1, EGL-19 and EXP-2 currents shape action potentials in the Caenorhabditis elegans pharynx. J Exp Biol 208:2177–2190

Singh B, Monteil A, Bidaud I, Sugimoto Y, Suzuki T, Hamano S, Oguni H, Osawa M, Alonso ME, Delgado-Escueta AV, Inoue Y, Yasui-Furukori N, Kaneko S, Lory P, Yamakawa K (2007) Mutational analysis of CACNA1G in idiopathic generalized epilepsy. Hum Mutat 28:524–525

Splawski I, Yoo DS, Stotz SC, Cherry A, Clapham DE, Keating MT (2006) CACNA1H mutations in autism spectrum disorders. J Biol Chem 281:22085–91

Strom SP, Stone JL, Ten Bosch JR, Merriman B, Cantor RM, Geschwind DH, Nelson SF (2010) High-density SNP association study of the 17q21 chromosomal region linked to autism identifies CACNA1G as a novel candidate gene. Mol Psychiatry 15:996–1005

Sui G, Fry CH, Malone-Lee J, Wu C (2009) Aberrant Ca^{2+} oscillations in smooth muscle cells from overactive human bladders. Cell Calcium 45:456–464

Talavera K, Nilius B (2006) Biophysics and structure-function relationship of T-type Ca^{2+} channels. Cell Calcium 40:97–114

Talley EM, Cribbs LL, Lee JH, Daud A, Perez-Reyes E, Bayliss DA (1999) Differential distribution of three members of a gene family encoding low voltage-activated (T-type) calcium channels. J Neurosci 19:1895–1911

Talley EM, Solorzano G, Depaulis A, Perez-Reyes E, Bayliss DA (2000) Low-voltage-activated calcium channel subunit expression in a genetic model of absence epilepsy in the rat. Mol Brain Res 75:159–165

Taylor JT, Huang L, Pottle JE, Liu K, Yang Y, Zeng X, Keyser BM, Agrawal KC, Hansen JB, Li M (2008a) Selective blockage of T-type Ca^{2+} channels suppresses human breast cancer cell proliferation. Cancer Lett 267:116–124

Taylor JT, Zeng XB, Pottle JE, Lee K, Wang AR, Yi SG, Scruggs JAS, Sikka SS, Li M (2008b) Calcium signaling and T-type calcium channels in cancer cell cycling. World J Gastroenterol 14:4984–4991

Toyota M, Ho C, Ohe-Toyota M, Baylin SB, Issa JP (1999) Inactivation of CACNA1G, a T-type calcium channel gene, by aberrant methylation of its 5' CpG island in human tumors. Cancer Res 59:4535–4541

Tringham E, Powell KL, Cain SM, Kuplast K, Mezeyova J, Weerapura M, Eduljee C, Jiang X, Smith P, Morrison JL, Jones NC, Braine E, Rind G, Fee-Maki M, Parker D, Pajouhesh H, Parmar M, O'Brien TJ, Snutch TP (2012) T-type calcium channel blockers that attenuate thalamic burst firing and suppress absence seizures. Sci Transl Med 4:121ra119

Uebele VN, Gotter AL, Nuss CE, Kraus RL, Doran SM, Garson SL, Reiss DR, Li Y, Barrow JC, Reger TS, Yang ZQ, Ballard JE, Tang C, Metzger JM, Wang SP, Koblan KS, Renger JJ (2009) Antagonism of T-type calcium channels inhibits high-fat diet-induced weight gain in mice. J Clin Invest 119:1659–1667

Ueki T, Toyota M, Sohn T, Yeo CJ, Issa JP, Hruban RH, Goggins M (2000) Hypermethylation of multiple genes in pancreatic adenocarcinoma. Cancer Res 60:1835–1839

Vassort G, Talavera K, Alvarez JL (2006) Role of T-type Ca^{2+} channels in the heart. Cell Calcium 40:205–220

Viscidi EW, Triche EW, Pescosolido MF, McLean RL, Joseph RM, Spence SJ, Morrow EM (2013) Clinical characteristics of children with autism spectrum disorder and co-occurring epilepsy. PLoS One 8:e67797

Vitko I, Chen Y, Arias JM, Shen Y, Wu XR, Perez-Reyes E (2005) Functional characterization and neuronal modeling of the effects of childhood absence epilepsy variants of CACNA1H, a T-type calcium channel. J Neurosci 25:4844–4855

Vitko I, Bidaud I, Arias JM, Mezghrani A, Lory P, Perez-Reyes E (2007) The I-II loop controls plasma membrane expression and gating of Ca(v)3.2 T-type Ca^{2+} channels: a paradigm for childhood absence epilepsy mutations. J Neurosci 27:322–330

Weiss N, Zamponi GW (2013) Control of low-threshold exocytosis by T-type calcium channels. Biochim Biophys Acta 1828:1579–1586

Weiss N, Hameed S, Fernández-Fernández JM, Fablet K, Karmazinova M, Poillot C, Proft J, Chen L, Bidaud I, Monteil A, Huc-Brandt S, Lacinova L, Lory P, Zamponi GW, De Waard M (2012) A Cav3.2/syntaxin-1A signaling complex controls T-type channel activity and low-threshold exocytosis. J Biol Chem 287:2810–2818

Weiss N, Black SA, Bladen C, Chen L, Zamponi GW (2013) Surface expression and function of Cav3.2 T-type calcium channels are controlled by asparagine-linked glycosylation. Pflugers Arch 465:1159–1170

Wen XJ, Xu SY, Chen ZX, Yang CX, Liang H, Li H (2010) The roles of T-type calcium channel in the development of neuropathic pain following chronic compression of rat dorsal root ganglia. Pharmacology 85:295–300

Wolfart J, Roeper J (2002) Selective coupling of T-type calcium channels to SK potassium channels prevents intrinsic bursting in dopaminergic midbrain neurons. J Neurosci 22:3404–3413

Yang J, Ellinor PT, Sather WA, Zhang JF, Tsien RW (1993) Molecular determinants of Ca^{2+} selectivity and ion permeation in L-type Ca^{2+} channels. Nature 366:158–161

Yeoman MS, Brezden BL, Benjamin PR (1999) LVA and HVA Ca^{2+} currents in ventricular muscle cells of the lymnaea heart. J Neurophysiol 82:2428–2440

Young RC, Smith LH, McLaren MD (1993) T-type and L-type calcium currents in freshly dispersed human uterine smooth muscle cells. Am J Obstetr Gynecol 169:785–792

Zamponi GW, Lewis RJ, Todorovic SM, Arneric SP, Snutch TP (2009) Role of voltage-gated calcium channels in ascending pain pathways. Brain Res Rev 60:84–89

Zhang Y, Mori M, Burgess DL, Noebels JL (2002) Mutations in high-voltage-activated calcium channel genes stimulate low-voltage-activated currents in mouse thalamic relay neurons. J Neurosci 22:6362–6371

Regulation of T-Type Ca^{2+} Channels by Intercellular and Intracellular Signals

2

Ming Li and Songwei Wu

Abstract

T-type Ca^{2+} channels play many important physiological functions in different tissues; this makes the channels targets for extracellular and intracellular regulation. T-type Ca^{2+} channels in many non-excitable cells provide an essential mechanism for Ca^{2+} entry at voltages near the resting membrane potential, which revises the homeostasis of the intracellular Ca^{2+} concentration to promote cell cycling and cell duplication. Consistent with this function, many of the extracellular signals stimulate the expression of T-type Ca^{2+} channels. The intracellular signaling pathways regulating the T-type Ca^{2+} channels are frequently observed in excitable cells. In these cases, the role of T-type Ca^{2+} current is to modify threshold and the shape of repetitive firing of action potentials. Therefore, intracellular signaling pathways can either up- or downregulate T-type Ca^{2+} channels. The regulation of the α_1H isoform of the T-type Ca^{2+} channels is frequently located on the intracellular loop between domains II and III.

2.1 Introduction

T-type Ca^{2+} channels are expressed throughout the body, including in nervous tissue, heart, kidney, smooth muscle, sperm, and endocrine glands. These channels have been implicated in diverse processes, including neuronal firing, hormone

M. Li (✉)
Department of Physiology, Tulane University Health Sciences Center, New Orleans, LA 70112, USA
e-mail: mli@tulane.edu

S. Wu
Department of Anesthesiology and Perioperative Medicine, Georgia Regents University, Augusta, GA 30912, USA

S.W. Schaffer, M. Li (eds.), *T-type Calcium Channels in Basic and Clinical Science*,
DOI 10.1007/978-3-7091-1413-1_2

secretion, smooth muscle contraction, myoblast fusion, and fertilization (Perez-Reyes 2003). Moreover, a noteworthy phenomenon of T-type Ca^{2+} channels that distinguish them from other voltage-gated Ca^{2+} channels is that they are commonly expressed in tissues undergoing proliferation. For example, T-type Ca^{2+} channels are generally expressed in fetal and neonatal heart, lung, and kidney (Wang et al. 1991; Gomez et al. 1994; Monteil et al. 2000), but they are downregulated and disappear in adult animals (Beam and Knudson 1988; Richard et al. 1992; Gomez et al. 1994; Sen and Smith 1994). Another example is that T-type Ca^{2+} channels have been found in malignant tumor tissues but not in their corresponding normal epithelia (Taylor et al. 2008b). Overexpression of the T-type Ca^{2+} channel doubles the proliferation rate of astrocytoma and neuroblastoma cells that already express the channels. Antisense oligonucleotides directed against mRNA encoding these channels reduce the proliferation rate but have no effect in cancer cells lacking channel expression (Panner et al. 2005).

T-type Ca^{2+} channels perform various functions in many types of cells, tissues, and organs and play a general role in cell proliferation. Therefore, they must be regulated vigorously by many intercellular and intracellular signals involved in both expression and activity. This chapter will review the regulation of T-type Ca^{2+} channels by some of the most important signal molecules, hormones, and intracellular signaling pathways.

2.2 Intercellular Signals

2.2.1 Glucose

Diseases related to fuel metabolism, including diabetes mellitus, obesity, hypertriglyceridemia, and hyperinsulinemia, all involve anomalies in plasma glucose regulation. Current data show that T-type Ca^{2+} channel expression is upregulated by high concentrations of glucose.

Infants develop hypertrophic cardiomyopathy in approximately 30 % of diabetic pregnancies. It has been found that the average T-type Ca^{2+} current density increases significantly in primary culture neonatal cardiac myocytes treated with 25 mM glucose for 48 h when compared to those treated with 5 mM glucose (Li et al. 2005). High-glucose treatment also causes higher cardiomyocyte Ca^{2+} influx in response to 50 mM KCl, an effect attenuated by nickel. Real-time PCR studies demonstrated that mRNA levels of both α_1G and α_1H T-type Ca^{2+} channels are elevated after high-glucose treatment. High-glucose also significantly increases ventricular cell proliferation, as well as the proportion of cells in S-phase of the cell cycle; both effects are reversed by nickel and mibefradil. These results indicate that high glucose causes a rise in $[Ca^{2+}]_i$ in neonatal cardiac myocytes by a mechanism related to the regulation of T-type Ca^{2+} channel gene expression.

In rat pancreatic insulin-secreting β-cells, T-type Ca^{2+} current density and mRNA (α_1G and α_1H) levels increase markedly after exposure to 11.1 mM glucose compared with 3.7 mM glucose treatment for 48 h (Keyser et al. 2014). In rat pancreatic islet cells chronically treated with high concentrations of glucose, $[Ca^{2+}]_i$ increased more than twofold, an effect which can be significantly inhibited by mibefradil (1 μM). Neither the L-type Ca^{2+} channel blocker nifedipine nor the Na^+–Ca^{2+} exchanger blocker benzamil exerted the same effect (Zhang et al. 2000a).

2.2.2 Cytokines

Inflammatory cytokines interferin-γ (IFN-γ), interleukin-1 (IL-1β), and tumor necrosis factor-α (TNF-α) play vital roles in many pathological processes, including the death of pancreatic β-cells in type 1 diabetes mellitus (Hamaguchi and Leiter 1990; Rbinovitch et al. 1994; Iwahashi et al. 1996; Dunger et al. 1997).

In mouse pancreatic islet cells, 48-h treatment with cytokines (300 U/ml INF-γ, 25 U/ml IL-1β) induced an expression of T-type Ca^{2+} channel currents (Wang et al. 1999). This current was present in 48 % of cytokine-treated cells. No such current was observed when cells were treated with either IL-β or IFN-γ alone or in non-treated cells. In contrast, the same cytokine cocktail failed to induce T-type Ca^{2+} channel current in a glucagon-secreting cell line α-TC1. Treatment of cytokines (IL-1β, INF-α, and 100 U/ml TNF-α) in mouse immortal β-cells, β-TC, or NIT cells (a β-cell line derived from NOD type 1 diabetic mouse model) also induced an augmentation of $[Ca^{2+}]_i$ and apoptotic cell death. However, neither alternation in basal $[Ca^{2+}]_i$ nor apoptotic cell death was observed in α-TC1 cells treated with the same cytokine cocktail. For the β-cells, this cytokine-induced apoptotic death can be blocked by concomitant incubation with T-type Ca^{2+} channel blockers, but not the L-type Ca^{2+} channel blocker nifedipine (Wang et al. 1999). This indicates that T-type Ca^{2+} channels play a critical role in switching β-cell status from living to death upon cytokine treatment.

Since cytokines are especially important in modulating the balance between humoral and cell-based immune responses as well as cell growth, maturation, and death, more research is needed to understand the role of T-type Ca^{2+} channels in cytokine-induced cell responses in a variety of diseases, including cancer.

Functional expression of T-type Ca^{2+} channels is developmentally regulated in chick nodose neurons (Pachuau and Martin-Caraballo 2007). Culturing E7 nodose neurons for 48 h with heart extract induces the expression of T-type Ca^{2+} channels with no significant effect on high-voltage-activated Ca^{2+} currents. Various hematopoietic cytokines, including ciliary neurotrophic factor (CNTF) and leukemia inhibitory factor (LIF), mimic the stimulatory effect of heart extract on T-type Ca^{2+} channel expression. A heterogeneous expression study showed that disruption of the Golgi apparatus with brefeldin A inhibited the stimulatory effect of LIF on $Ca_v3.2$ (α_1H) T-type Ca^{2+} channels, suggesting that protein trafficking regulates the functional expression of T-type Ca^{2+} channels (Dey et al. 2011).

Macrophage migration inhibitory factor (MIF) downregulates T-type Ca^{2+} current in atrial-derived HL-1 cells (Rao et al. 2013). The inhibitory effect of MIF is mediated by reducing the transcription of α_1G and α_1H, as well as shifting the activation curve toward depolarizing potentials. The mechanism of this inhibitory effect may involve activation of Src family tyrosine kinase since genistein and PP1 both significantly restore the level of T-type Ca^{2+} channel transcription and current amplitude (Rao et al. 2013).

2.2.3 Hormones

2.2.3.1 Testosterone

The general role of T-type Ca^{2+} channels in regulating cell proliferation makes these proteins especially interesting in controlling tumor growth. Indeed, T-type Ca^{2+} channels have been reported to play a vital role in the progress of many types of tumors (Taylor et al. 2008b). Since the growth of some tumors is hormone driven or associated, it is important to disclose the effect of hormones on the regulation of T-type Ca^{2+} channels.

Androgens (testosterone and dihydrotestosterone) stimulate prostate cancer cells to grow. Lowering androgen levels or blocking their effects often promotes prostate cancer shrinkage or causes slower growth. Human prostate cancer cells express T-type Ca^{2+}channels [Gray et al. 2004; Gackière et al. 2008). Blockade of these channels decreases prostate cancer cell proliferation (Haverstick et al. 2000).

The regulation of T-type Ca^{2+} channels by testosterone has been studied in neonatal rat cardiomyocytes (Michels et al. 2006). Testosterone pretreatment significantly increases whole-cell T-type Ca^{2+} channel current density and accelerates cellular beating rate. The increase in T-type Ca^{2+} channel current is related to the upregulation of both α_1G and α_1H subunits and an elevation in the opening probability of the channels. In contrast, acute application of testosterone at high concentrations decreases T-type Ca^{2+} channel current (Michels et al. 2006).

2.2.3.2 Estrogen

Estrogens play an important role in the regulation of growth, differentiation, and function in a wide array of target tissues, including breast and ovarian cancers (Taylor et al. 2008a; Li et al. 2011; Pottle et al. 2013). In both cases, selective inhibition of T-type Ca^{2+} channels by siRNA targeting of α_1G and α_1H subunits significantly inhibits cancer cell proliferation.

The effects of estrogen (17β-estradiol) on calcium current density were examined in GH_3 anterior pituitary cells (Ritchie 1993). Chronic treatment of these cells with estrogen induced a four- to fivefold increase in T-type Ca^{2+} current, whereas in the same experiments estrogen had no effect on the regulation of high-voltage-activated Ca^{2+} current. No change was observed in the voltage dependence or the time course of activation, nor in inactivation or deactivation of the current. The change in the size of T-type Ca^{2+} current was associated with an increase in the number of functional proteins on the cytoplasmic membrane, which was blocked by the protein synthesis inhibitor cycloheximide.

Like testosterone, estrogen downregulates the actions of the T-type Ca^{2+} channel. In cardiomyocytes, estrogen treatment for 24 h caused an estrogen receptor-independent downregulation of the peak current of the T-type Ca^{2+} channel (Marni et al. 2009). Estrogen also suppressed mRNA expression of α_1H but not α_1G. The inhibitory effect of estrogen has been linked to cytoplasmic ERK-1/2, since PD-98059 abolishes the effects of estrogen on T-type Ca^{2+} currents.

The exact reason for these dual effects of testosterone and estrogen on the regulation of T-type Ca^{2+} channels is unclear. One explanation is that long-term treatment with these hormones helps establish an intracellular calcium environment that is beneficial for cell growth or proliferation. However, the changes in the excitability of cardiomyocytes may be an "unwanted" effect, which could be reduced by inhibiting the T-type Ca^{2+} channel via different pathways.

2.2.3.3 Aldosterone/Angiotensin II

The aldosterone/angiotensin system plays a critical role in regulating body blood pressure. T-type Ca^{2+} channels also play an important role in blood pressure regulation, but what effect does the aldosterone/angiotensin system have on the regulation of T-type Ca^{2+} channels?

In adrenal H295R cells, aldosterone treatment selectively increases T-type Ca^{2+} current density without affecting L-type Ca^{2+} channel current. This upregulation involves a 40 % increase in α_1H messenger level in response to aldosterone treatment; however, α_1G is insensitive to the treatment (Rossier et al. 2003). In rat neonatal cardiomyocytes, aldosterone increases α_1H expression but decreases α_1G expression (Rossier et al. 2003). Aldosterone also increases α_1H expression in prostatic LNCaP cells (Rossier et al. 2003).

The secretion of aldosterone by adrenal glomerulosa cells is stimulated by angiotensin II, a process involving an increase in T-type Ca^{2+} current (Cohen et al. 1988). Angiotensin II stimulates α_1G and α_1H expression in cardiomyocytes through a signaling pathway involving extracellular signal-regulated kinase (ERK) 1/2 and p38 mitogen-activated protein kinase (MAPK) phosphorylation (Ferron et al. 2003; Morishima et al. 2009). The increase in T-type Ca^{2+} current density by angiotensin II observed in cardiomyocytes may play a role in the pathogenesis of Ca^{2+} overload and arrhythmias seen in cardiac pathology.

2.2.3.4 IGF-I

The effect of insulin like growth factor-I (IGF-I) on the regulation of T-type Ca^{2+} channels has been examined in pulmonary artery smooth muscle cells (PASMCs). IGF-I upregulates α_1G mRNA via a phosphatidylinositol 3-kinase/Akt signaling pathway (Plueanu and Cribbs 2011). The expression of the T-type Ca^{2+} channels plays a critical role in elevating $[Ca^{2+}]_i$, which in turn leads to cell cycle initiation and proliferation of PASMCs. Knocking down α_1G expression by RNA interference inhibits the expression and activation of cyclin D, which is necessary for cell cycle progression.

The proliferation of PASMCs is a key factor in the pathogenesis of diseases that involve vessel remodeling. IGF-I is produced both in vascular endothelial cells and smooth muscle cells. The effect of IGF-I in regulating the T-type Ca^{2+} channel suggests that regulation of these channels may play a critical role in the development of pathological conditions in the heart and vascular system. The definite function of T-type Ca^{2+} channels in controlling IGF-I-induced cell proliferation in PASMCs may also be present in the mechanism of tumor cell proliferation.

2.2.4 Anandamide

Endocannabinoids serve as intercellular "lipid messengers," signaling molecules that are released from one cell and activate cannabinoid receptors CB1/CB2 present on nearby cells. It has been reported that anandamide at submicromolar concentrations directly inhibits α_1G, α_1H, and α_1I T-type Ca^{2+} channels exogenously expressed in HEK 293 cells, as well as native T-type Ca^{2+} current of NG108-15 cells (Chemin et al. 2001). The inhibitory effect of anandamide is blocked by the anandamide membrane transport inhibitor AM404, suggesting that anandamide acts on the intracellular domains of the T-type Ca^{2+} channel. The mechanism of inhibition was also revealed: Anandamide binds and stabilizes T-type Ca^{2+} channels in their inactive state and reduces their activity. In CB1 and CB2 neurons, the probable function of anandamide is to modulate the excitability of the cells. Interestingly, anandamide has been shown to inhibit human breast cancer cell proliferation (De Petrocellis et al. 1998). This effect can also be explained by direct inhibition of the T-type Ca^{2+} channels of breast cancer cells.

Endogenous anandamide has a very short half-life, due to its action on the enzyme fatty acid amide hydrolase (FAAH), which breaks it down into free arachidonic acid and ethanolamine. Arachidonic acid, like anandamide, also attenuates T-type Ca^{2+} current. The application of 10 μM arachidonic acid causes a slow, time-dependent reduction in α_1H Ca^{2+} current of HEK 293 cells (Zhang et al. 2000b). Arachidonic acid may exert its vasodilation effect in part by reducing calcium influx mediated by T-type Ca^{2+} channels in vascular smooth muscle cells.

Metabolites of arachidonic acid, epoxyeicosatrienoic acids (EETs), also potently inhibit T-type Ca^{2+} channels (Cazade et al. 2013). A P450 epoxygenase product, 5,6-EET, inhibits $Ca_V3.1$ (α_1G), $Ca_V3.2$ (α_1H), and $Ca_V3.3$ (α_1H) channels heterogeneously expressed in tsA-201 cells with an IC_{50} of 0.54 μM.

2.2.5 Nitric Oxide/Nitrous Oxide

Nitric oxide (NO) is an important cell-signaling molecule involved in many physiological and pathological processes. It is a powerful vasodilator with a short half-life of a few seconds in the blood. In human cavernosal smooth muscle cells, NO mediates an immediate decrease in $[Ca^{2+}]_i$, which is responsible for muscle relaxation and vasodilation (Lue 2000). This pathway (presented as pathway 1 in Fig. 2.1) involves the opening of K^+ channels, which in turn causes membrane hyperpolarization and the closing of L-type Ca^{2+} channels. Moreover, NO mediates a delayed increase in $[Ca^{2+}]_i$ as reflected in pathway 2 of Fig. 2.1. This resulted from an increase in the expression of T-type (α_1G) Ca^{2+} channels (Zeng et al. 2005). In

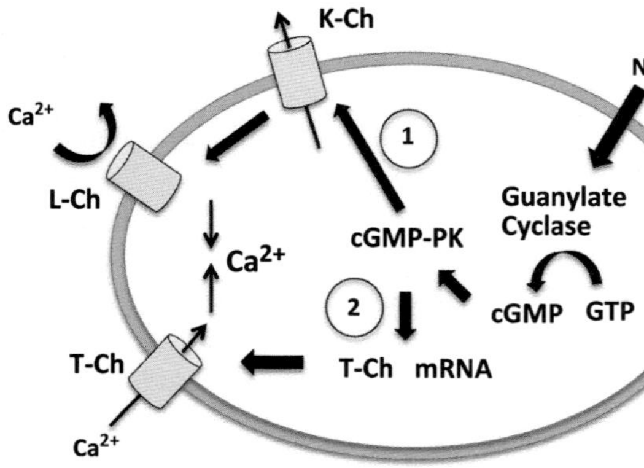

Fig. 2.1 A scheme showing T-type Ca^{2+} channels with NO-cGMP pathway in HCSMC. NO-induced transient decrease in $[Ca^{2+}]_i$ by opening K^+ channels and closing L-type Ca^{2+} channels and maintains Ca^{2+} homeostasis by increased expression of T-channels via upregulating mRNA transcription, a delayed effect of NO-cGMP-induced cG-PK activation

cavernosal smooth muscle cells, the second pathway is blocked by the selective T-type Ca^{2+} channel blocker NNC 55-0396. Pretreatment of NO does not alter mRNA levels of L-type Ca^{2+} channels, suggesting that NO elevates $[Ca^{2+}]_i$ via an induction of T-type Ca^{2+} channel expression (Zeng et al. 2005). Physiologically, the upregulation of T-type Ca^{2+} channel and consequent elevation of $[Ca^{2+}]_i$ contributes to the development of insensitivity to phosphodiesterase type 5 (PDE5) inhibitors. This mechanism may also be involved in NO-stimulated smooth muscle cell proliferation (Ignarro et al. 2001).

The acute effect of NO-cGMP-PKG signaling on T-type Ca^{2+} currents has been examined in the rat cerebral arterial smooth muscle cells (Harraz et al. 2014). NO donors or activators of protein kinase G suppress T-type Ca^{2+} current measured by patch clamp. However, the subtypes of T-type Ca^{2+} channels involved in this downregulation are yet to be identified.

The effect of nitrous oxide (N_2O) on the level of T-type Ca^{2+} current has been examined in rat neurons. At subanesthetic concentrations, N_2O blocks T-type Ca^{2+} current without affecting high-voltage activated Ca^{2+} current (Todorovic et al. 2001). Interestingly, N_2O also selectively blocks α_1H but not the α_1G subtype T-type Ca^{2+} channels, which are heterogeneously expressed in HEK 293 cells. Blockade of the T-type Ca^{2+} current may contribute to the anesthetic and/or analgesic effects of N_2O.

2.2.6 Zinc

Released from the presynaptic vesicles of glutamatergic neurons, zinc differentially regulates the three T-channel subtypes. In the submicromolar range (~0.8 μM), zinc inhibits the $Ca_v3.2$ (α_1H) isoform of the T-type Ca^{2+} channels, but inhibition of the

$Ca_v3.1$ (α_1G) or $Ca_v3.3$ (α_1I) isoforms requires a 100- to 200-fold increase in zinc concentration (Traboulsie et al. 2007). Zinc also causes a slow deactivation of the tail current of the T-type Ca^{2+} channel (Traboulsie et al. 2007). The reducing agent L-cysteine was reported to restore T-type Ca^{2+} current by chelation of free zinc ions (Nelson et al. 2007a). The inhibitory effect of zinc on the T-type Ca^{2+} channel has been pinpointed to the extracellular histidine (His191) residue localized in the S3–S4 segment of domain I of the $Ca_v3.2$ (α_1H) T-type Ca^{2+} channel (Nelson et al. 2007b). Residue His191 is also the binding site for nickel, a broadly known pharmacological blocker of the T-type Ca^{2+} channel (Kang et al. 2006). Ascorbate also interacts with His191 to inhibit the T-type Ca^{2+} channel via metal-catalyzed oxidation (Nelson et al. 2007a).

2.3 Intracellular Signals

2.3.1 Heterotrimeric G-Protein

Malfunction of the GPCR (G-protein-coupled receptor) signaling pathways is involved in many health problems, including diabetes, blindness, allergies, depression, cardiovascular defects, and certain forms of cancer. It is estimated that about 30 % of the modern drugs' cellular targets are GPCRs (Bosch and Siderovski 2013). The most G-protein-regulated T-type Ca^{2+} channel isoform is $Ca_v3.2$ (α_1H). This type of isoform can be directly inhibited by the $\beta_2\gamma_2$ subunit dimer of G-proteins in response to dopamine D1 stimulation without a change in channel voltage dependency (Wolfe et al. 2003; DePuy et al. 2006). The channel-regulating site that interacts with $\beta_2\gamma_2$ subunits of the G-protein is located on the intracellular loop between domain II and III (DePuy et al. 2006). In contrast, a study involving corticotropin-releasing factor receptor 1 (CRFR1) inhibits T-type Ca^{2+} channels, demonstrating that the $\beta_2\gamma_2$ subunits exert an inhibitory effect on the α_1H T-type Ca^{2+} channel, probably by stabilizing the channels in the inactive state (Tao et al. 2008).

Activation of the muscarinic M1 receptor has a strong inhibitory effect on $Ca_v3.3$ (α_1I) T-type Ca^{2+} currents but has either no effect or a moderate stimulating effect on peak current amplitude of $Ca_v3.1$ (α_1G) and $Ca_v3.2$ (α_1H) (Hildebrand et al. 2007). In contrast to the CRFR1 pathway, the M1 receptor-mediated inhibition of the $Ca_v3.3$ (α_1I) T-type Ca^{2+} channels occurs through $G\alpha_{q/11}$ (Hildebrand et al. 2007).

2.3.2 Rho-Associated Kinase

The regulation of T-type Ca^{2+}channels by Rho-associated kinase (ROCK) has been investigated in HEK cells expressing transfected isoforms of the channels. Activation of ROCK via the endogenous ligand lysophosphatidic acid (LPA) diminishes peak currents of rat $Ca_v3.1$ (α_1G) and $Ca_v3.3$ (α_1I) T-type Ca^{2+}channels without affecting the voltage dependence of activation and inactivation (Iftinca et al. 2007). LPA also reduces peak current of $Ca_v3.2$ (α_1H) as well as shifts the activation and inactivation curves in a more positive direction. Site-directed mutagenesis of the α_1G subunit reveals that the ROCK-mediated effects involve two distinct phosphorylation consensus sites in the intracellular loop between domain II and III of the subunit (Iftinca et al. 2007).

The effects of LPA on native T-type Ca^{2+} current are cell type dependent; in Y79 retinoblastoma and in lateral habenular neurons, activation of LPA causes a reduction in native T-type Ca^{2+} current, whereas in dorsal root ganglion neurons, it potentiates native α_1H current. The mechanism underlying this upregulation is unclear but may involve a positive shift in voltage-dependent channel activation (Iftinca et al. 2007).

2.3.3 Protein Kinase C

Studies of protein kinase C (PKC)-dependent regulation of T-type Ca^{2+} channels have generated rather complicated results. The activation of PKC downregulates T-type Ca^{2+} current in rat DRG neurons (Schroeder et al. 1990), canine Purkinje cells (Tseng and Boyden 1991), NIH-3T3 fibroblast cells (Pemberton et al. 2000), and GH_3 cells (Marchetti and Brown 1988). However, PKC potentiates T-type Ca^{2+} current in both neonatal rat ventricular myocytes (Furukawa et al. 1992) and of α_1H T-type Ca^{2+} channels reconstituted in *Xenopus* oocytes (Park et al. 2006). The current of recombinant $Ca_v3.2$ (α_1H) subunits of the T-type Ca^{2+} channel is reversibly inhibited by active neurokinin 1 (NK1) receptors when both proteins are transiently expressed in HEK293 cells. This inhibition was blocked by a selective inhibitor of PKC, bisindolylmaleimide I (Rangel et al. 2010).

The precise mechanism underlying the differential modulation of T-type Ca^{2+} channel activity by PKC is unclear. Some of the discrepancy can be attributed to the concentration difference of the PKC activator used in the experiments. For example, in reconstituted *Xenopus* oocytes, 200 nM PMA enhances α_1H T-type Ca^{2+} current amplitude whereas 10 and 100 nM PMA inhibits the current (Park et al. 2003).

A multiplicity of functions has been ascribed to PKC, which is regulated by two of the major second messengers of cells (diacylglycerol and Ca^{2+}). Recurring themes are that PKC is involved in a variety of intracellular functions, including receptor desensitization, transcription regulation and cell growth, learning, and memory. From the multiplicity of functions of PKC, we can ascertain that its effects are cell type specific. It is understandable that the regulatory effect of PKC on T-type Ca^{2+} channels may also be different depending on the preparations. For example, effects of PKC on different isoforms of T-type Ca^{2+} channels can vary (Shan et al. 2013).

2.3.4 ATP

In thalamocortical neurons, intracellular perfusion of ATP potentiates T-type Ca^{2+} current (Leresche et al. 2004). This potentiation was prevented if the pipette solution lacks ATP and becomes irreversible if ATP is replaced by the nonhydrolyzable analog ATPγS. This result suggests that the potentiating effect of ATP on the amplitude of T-type Ca^{2+} current is likely phosphorylation dependent.

Paradoxically, phosphorylation also stabilizes the channel inactivation kinetics, which results in a reduction in current amplitude. Hence, phosphorylation generates two opposite effects on T-type Ca^{2+} current in these neurons. The net effect is time dependent: After transient hyperpolarization, the de-inactivation and potentiation effects dominate resulting in maximal current. After more prolonged hyperpolarization, the phosphorylation and inactivation effects are dominant, resulting in greater inhibition of T-type Ca^{2+} current. Therefore, the ATP/phosphorylation pathway fine-tunes T-type Ca^{2+} channel activity and thus neuronal excitability.

2.3.5 cAMP and Protein Kinase A

T-type Ca^{2+} channels are weakly expressed or virtually absent in adult rat chromaffin cells; however, intense T-type Ca^{2+} channel current appears after application of pCPT-cAMP in serum-free culture media for 3–5 days (Novara et al. 2004). A similar effect is obtained when replacing cAMP with forskolin, 1-methyl-3-isobutylxanthine (IBMX), or isoprenaline. In contrast, these treatments have no effect on current density of the high-voltage-activated Ca^{2+} channels. The action of cAMP is prevented by the protein synthesis inhibitor anisomycin, suggesting that the effect involves the synthesis of T-type Ca^{2+} channels. Channel recruitment is not blocked by H89 or Rp-cAMP but is mimicked by the selective cAMP-activated guanine nucleotide exchange factors for Ras-like GTPases (Epac) agonist, 8CPT-2Me-cAMP, suggesting the existence of a prominent Epac-dependent T-type Ca^{2+} channel recruitment mechanism in response to β-adrenergic stimulation.

The cAMP-Epac signaling pathway is also involved in the upregulation of T-type Ca^{2+} channels in rat chromaffin cells during chronic hypoxia. Exposure of chromaffin cells to 3 % O_2 for 12–18 h causes comparable expression of α_1H T-type Ca^{2+} current, which plays a role in low-voltage exocytosis (Carabelli et al. 2007).

Protein kinase A (PKA) upregulates the activity of $Ca_v3.2$ (α_1H) T-type channels reconstituted in *Xenopus* oocytes. The application of forskolin increases $Ca_v3.2$ (α_1H) channel activity and negatively shifts the steady-state inactivation curve. The chimeric channels constructed by replacing the cytoplasmic intracellular loops with corresponding segments of the $Na_v1.4$ Na^+ channel (which are insensitive to PKA) loses their sensitivity to PKA, indicating that the II–III loop contains structural element(s) critical to PKA stimulation (Kim et al. 2006).

PKA downregulates the T-type Ca^{2+} channels in rat cerebral arterial smooth muscle cells. In one study, PKA modulators, db-cAMP, forskolin or isoproterenol, suppress α_1H T-type Ca^{2+} current (Harraz and Welsh 2013). This inhibition is associated with a hyperpolarizing shift in steady-state inactivation. It has been suggested that vasodilatory signaling cascades inhibit T-type Ca^{2+} channels via PKA activation, thereby reducing Ca^{2+} influx and regulating cerebral circulation.

PKA also plays a permissive role in G-protein-mediated T-type Ca^{2+} channel regulation. As discussed previously, activation of GPCR inhibits α_1H T-type Ca^{2+} channel activity by $G\beta_2\gamma$ dimer (Wolfe et al. 2003). This inhibition requires phosphorylation of Ser1107 on the II–III loop of the channel pore protein. $G\beta_2\gamma$ dimers released upon receptor activation also require PKA activity for their inhibitory actions. Since dopamine inhibition of $Ca_v3.2$ (α_1H) whole-cell current is precluded by blocking PKA catalytic activity, PKA serves as a molecular switch that allows $G\beta_2\gamma$ dimers to regulate $Ca_v3.2$ (α_1H) channels (Hu et al. 2009).

2.3.6 Ca^{2+} and Ca^{2+}/Calmodulin-Dependent Kinase II

Intracellular free calcium itself directly regulates T-type Ca^{2+} channel activity. The "high" $[Ca^{2+}]_i$ potentiates T-type Ca^{2+} current, as indicated by the increase in current at −20 mV and lower voltages, as well as the tail current measured in pancreatic β-cells (Huang et al. 2004). The voltage-dependent activation curve of T-type Ca^{2+} channel current is shifted ~11 mV to negative potentials after "high" Ca^{2+} perfusion.

Ca^{2+}/Calmodulin-dependent kinase II (CaMKII) forms a signaling complex with α_1H T-type Ca^{2+} channels, directly interacting with the intracellular loop connecting domains II and III of the channel pore (Welsby et al. 2003; Yao et al. 2006). Activation of the kinase leads to phosphorylation of Ser1198 in the II–III loop and causes a hyperpolarizing shift of α_1H T-type Ca^{2+} channel activation kinetics both in intact cells in situ and in cells of the native adrenal zona glomerulosa stimulated by angiotensin II in vivo. Thus, α_1H channels use tethered CaMKII as a Ca^{2+} sensor to dynamically regulate channel activity and Ca^{2+} entry in angiotensin II-targeting cells.

2.3.7 Tyrosine Kinase

The effect of tyrosine kinase on T-type Ca^{2+} channel activity has been studied using the $Ca_v3.2$ (α_1H) isoform reconstituted in *Xenopus* oocytes (Sun-Un et al. 2008). $Ca_v3.2$ (α_1H) channel activity is reduced by 25 % in response to phenylarsine oxide (tyrosine phosphatase inhibitor), whereas it is augmented by 19 % in response to Tyr A47 or herbimycin A (tyrosine kinase inhibitors). However, other biophysical properties of $Ca_v3.2$ (α_1H) current are not significantly changed by the drugs. These results imply that $Ca_v3.2$ (α_1H) T-type Ca^{2+} channel activity is increased by activation of tyrosine phosphatases, but decreased by activation of the tyrosine kinases.

Cultured chicken nodose neurons exhibit increased T-type Ca^{2+} channel expression in response to CNTF (Pachuau and Martin-Caraballo 2007). This expression is not affected by inhibition of protein synthesis, suggesting the involvement of posttranslational regulation (Pachuau and Martin-Caraballo 2007). The expression of T-type Ca^{2+} channels was blocked by the Janus tyrosine kinase (JAK) inhibitor P6 or the extracellular signal-regulated kinase (ERK) inhibitor U0126 (Trimarchi et al. 2009). However, inhibition of STAT3 with static has no effect on T-type Ca^{2+} channel expression. This suggests that CNTF-evoked stimulation of T-type Ca^{2+} channel expression requires JAK-dependent ERK signaling.

2.3.8 Transcription Factors

Changes in T-type Ca^{2+} channel expression have been observed during organ development, as well as under various pathological states. However, little is known about the transcriptional regulation of T-type Ca^{2+} channel gene expression. In humans, the genes coding α_1H and α_1G are located on chromosomes 16 and 17, respectively. Two 5'-flanking regions of cloned α_1G (Toyota et al. 1999) lack TATA and CAAT boxes, a characteristic commonly shared by so-called housekeeping genes. These two previously described alternative promoters have been used to generate short (differentiated cell) or long (undifferentiated cell) transcripts (Bertolesi et al. 2003). Furthermore, this gene is subjected to extensive alternative splicing (11 sites) including two alternative promoters, as well as two alternative polyadenylation sites (Latour et al. 2004; Emerick et al. 2006).

Cacna1g encoding α_1G T-type Ca^{2+} channel protein is a large gene that contains 38 exons and several potential transcription factor binding motifs (Perez-Reyes 2003). A study using the mouse neuroblastoma NIE-115 cell line characterized a 5' upstream region of *Cacna1g* and found that this region possesses a TATA-less and high-GC content minimal promoter that includes two potential transcription start sites and four binding sites for the transcription factor Sp1 (González-Ramírez et al. 2014). Overexpression of Sp1 enhances promoter activity of *Cacna1g* (González-Ramírez et al. 2014). Sp1 is an ubiquitously expressed zinc finger-containing DNA binding protein that can target gene products, including factors involved in the cell cycle, angiogenesis, apoptosis, and regulation such as proto-oncogenes and tumor suppressors (Suske 1999; Wuerstra 2008; Li and Davie 2010). This may account for the general role of T-type Ca^{2+} channels in cell proliferation.

Overexpression of a cardiac-specific transcription factor Csx/Nkx2.5 markedly increases expression of $Ca_v3.2$ (α_1H) T-type Ca^{2+} channels in neonatal rat hearts (Wang et al. 2007). In contrast, expression of Csx/Nkx2.5 downregulates L-type Ca^{2+} channels ($Ca_v1.3$) expression. In the same study, transcription factor GATA4 was found to exert no effect on T-type Ca^{2+} channel expression. The inactivation of Csx/Nks2.5 is also involved in the effects of estrogen on α_1H, but not α_1G, T-type Ca^{2+} channels in cardiomyocytes (Marni et al. 2009).

Conclusions

T-type Ca^{2+} channels can be regulated at both the transcriptional and posttranslational level. Transcriptional regulation is likely related to cell proliferation or apoptosis, processes mostly involving upregulation of T-type Ca^{2+} channel expression. This observation indicates that T-type Ca^{2+} channels in many non-excitable cells provide an essential mechanism of Ca^{2+} entry, which revises the homeostasis of intracellular Ca^{2+} concentration to promote cell cycling and cell duplication. For instance, increased expression of T-type Ca^{2+} channels seen in embryonic and neonatal tissues reflects changes that help facilitate cell proliferation. The specific electrophysiological characteristics of the T-type Ca^{2+} channels, such as window current at low membrane potentials and slow decaying tail currents, likely provide unique modes for Ca^{2+} entry at resting membrane potentials.

The posttranslational regulations of T-type Ca^{2+} channels are frequently observed in excitable cells. In these cases, the role of T-type Ca^{2+} currents is to modify threshold and the shape of action potentials. Therefore, intracellular signaling pathways can either up- or downregulate T-type Ca^{2+} channels. Many intracellular signaling pathways exert their regulatory effects specifically on the intracellular loop between domains II and III of the α_1H isoform of the T-type Ca^{2+} channels. This phenomenon suggests that α_1H plays a vital role in modifying the excitability of neurons and myocytes whereas α_1G is preferentially involved in regulating basal $[Ca^{2+}]_i$ in proliferating cells.

References

Beam KG, Knudson CM (1988) Effect of postnatal development on calcium currents and slow charge movement in mammalian skeletal muscle. J Gen Physiol 91:799–815

Bertolesi GE, Jollimore CA, Shi C, Elbaum L, Denovan-Wright EM, Barnes S, Kelly ME (2003) Regulation of α1G T-type calcium channel gene (CACNA1G) expression during neuronal differentiation. Eur J Neurosci 17:1802–1810

Bosch DE, Siderovski DP (2013) G protein signaling in the parasite *Entamoeba* histolytica. Exp Mol Med 10:1–12

Carabelli V, Marcantoni A, Comunanza V, de Luca A, Díaz J, Borges R, Carbone E (2007) Chronic hypoxia up-regulates α1H T-type channels and low-threshold catecholamine secretion in rat chromaffin cells. J Physiol 584:149–165

Cazade M, Bidaud I, Hansen PB, Lory P, Chemin J (2013) 5,6-EET potently inhibits T-type calcium channels: implication in the regulation of the vascular tone. Pflugers Arch. doi:10.1007/s00424-013-1411-0

Chemin J, Monteil A, Perez-Reyes E, Nargeot J, Lory P (2001) Direct inhibition of T-type calcium channels by the endogenous cannabinoid anandamide. EMBO J 20:7033–7040

Cohen CJ, McCarthy RT, Barrett PQ, Rasmussen H (1988) Ca channels in adrenal glomerulosa cells: K^+ and angiotensin II increase T-type Ca channel current. Proc Natl Acad Sci USA 85:2412–2416

De Petrocellis L, Melck D, Palmisano A, Bisogno T, Laezza C, Bifulco M, Di Marzo V (1998) The endogenous cannabinoid anandamide inhibits human breast cancer cell proliferation. Proc Natl Acad Sci USA 95:8375–8380

DePuy SD, Yao J, Hu C, McIntire W, Biduad I, Lory P, Rastinejad F, Gonzalez C, Garrison JC, Barrett PQ (2006) The molecular basis for T-type Ca^{2+} channel inhibition by G protein $\beta_2\gamma_2$ subunits. Proc Natl Acad Sci USA 103:14590–14595

Dey D, Shepherd A, Pachuau J, Martin-Caraballo M (2011) Leukemia inhibitory factor regulates trafficking of T-type Ca^{2+} channels. Am J Physiol 300:C576–C587

Dunger A, Cunningham JM, Delaney CA, Pipeleers DG, Lowe JE, Green MHL, Bone AJ, Green IC (1997) Cytokine induced deoxyribonucleic acid strand breaks and apoptosis in human pancreatic islet cells. Endocrinology 183:2610–2614

Emerick MC, Stein R, Kunze R, McNulty MM, Regan MR, Hanck DA, Agnew WS (2006) Profiling the array of $Ca_V3.1$ variants from the human T-type calcium channel gene CACNA1G: alternative structures, developmental expression, and biophysical variations. Proteins 64:320–342

Ferron L, Capuano V, Ruchon Y, Deroubaix E, Coulombe A, Renaud JF (2003) Angiotensin II signaling pathways mediate expression of cardiac T-type calcium channels. Circ Res 93:1241–1248

Furukawa T, Ito H, Nitta J, Tsujino M, Adachi S, Hiroe M, Marumo F, Sawanobori T, Hiraoka M (1992) Endothelin-1 enhances calcium entry through T-type calcium channels in cultured neonatal ventricular myocytes. Circ Res 71:1242–1253

Gackière F, Bidaux G, Delcourt P, Van Coppenolle F, Katsogiannou M, Dewailly E, Bavencoffe A, Tran Van Chuoï-Mariot M, Mauroy B, Prevarskaya N, Mariot P (2008) $Ca_V3.2$ T-type calcium channels are involved in calcium-dependent secretion of neuroendocrine prostate cancer cells. J Biol Chem 283:10162–10173

Gomez JP, Potreau D, Branka JE, Raymond G (1994) Developmental changes in Ca^{2+} current from newborn rat cardiomyocytes in primary culture. Pflüger Arch 428:241–249

González-Ramírez R, Martínez-Hernández E, Sandoval A, Felix R (2014) Transcription factor Sp1 regulates T-type Ca^{2+} channel $Ca_v3.1$ gene expression. J Cell Physiol 229:551–560

Gray LS, Perez-Reyes E, Gomora JC, Haverstick DM, Shattock M, McLatchie L, Harper J, Brooks G, Heady T, MacDonald TL (2004) The role of voltage gated T-type Ca^{2+} channel isoforms in mediating “capacitative” Ca^{2+} entry in cancer cells. Cell Calcium 36:489–497

Hamaguchi K, Leiter EH (1990) Comparison of cytokine effects on mouse pancreatic α-cell and β-cell line: viability, secretory function, and MHC antigen expression. Diabetes 39:415–425

Harraz OF, Welsh DG (2013) Protein kinase A regulation of T-type Ca^{2+} channels in rat cerebral arterial smooth muscle. J Cell Sci 126:2944–2954

Harraz OF, Brett SE, Welsh DG (2014) Nitric oxide suppresses vascular voltage-gated T-type Ca^{2+} channels through cGMP/PKG signaling. Am J Physiol 306:H279–H285

Haverstick DM, Heady TN, Macdonald TL, Gray LS (2000) Inhibition of human prostate cancer proliferation in vitro and in a mouse model by a compound synthesized to block Ca^{2+} entry. Cancer Res 60:1002–1008

Hildebrand ME, David LS, Hamid J, Mulatz K, Garcia E, Gerald W, Zamponi GW, Snutch TP (2007) Selective inhibition of Cav3.3 T-type calcium channels by Gαq/11-coupled muscarinic acetylcholine receptors. J Biol Chem 282:21043–21055

Hu C, DePuy SD, Yao J, McIntire WE, Barrett PQ (2009) Protein kinase A activity controls the regulation of T-type $Ca_v3.2$ channels by $G_{\beta\gamma}$ dimers. J Biol Chem 284:7465–7473

Huang L, Bhattacharjee A, Taylor JT, Zhang M, Keyser BM, Marrero L, Li M (2004) $[Ca^{2+}]_i$ regulated $Ca_V1.3$ translocation in insulin secreting cells. Am J Physiol 286:C213–C221

Iftinca M, Jawed Hamid J, Chen L, Varela D, Tadayonnejad R, Altier C, Ray W, Turner RW, Zamponi GW (2007) Regulation of T-type calcium channels by Rho-associated kinase. Nat Neurosci 7:854–860

Ignarro LJ, Buga GM, Wei LH, Bauer PM, Wu G, del Soldato P (2001) Role of the arginine-nitric oxide pathway in the regulation of vascular smooth muscle cell proliferation. Proc Natl Acad Sci USA 98:4202–4208

Iwahashi H, Hanafusa T, Eguchi Y, Nakajima H, Miyagawa J, Itoh N, Tomita K, Namba M, Kuwajima M, Noguchi T, Tsujimoto Y, Matzuzawa Y (1996) Cytokine-induced apoptotic cell death in a mouse pancreatic beta-cell line: inhibition by Bcl-2. Diabetologia 39:530–536

Kang HW, Park JY, Jeong SW, Kim JA, Moon HJ, Perez-Reyes E, Lee JH (2006) A molecular determinant of nickel inhibition in Cav3.2 T-type calcium channels. J Biol Chem 281:4823–4830

Keyser BM, Taylor JT, Choi S-K, Lu Y, Bhattacharjee A, Huang L, Pottle J, Matrougui K, Xu Z, Li M (2014) Role of T-type Ca^{2+} channels in basal $[Ca^{2+}]_i$ regulation and basal insulin secretion in rat islet cells. Curr Trend Endocrinol 7:35–44

Kim J-A, Park J-Y, Kang H-W, Huh S-U, Jeong S-W, Lee J-H (2006) Augmentation of $Ca_v3.2$ T-Type Calcium Channel Activity by cAMP-Dependent Protein Kinase A. J Pharmacol Exp Ther 318:230–237

Latour I, Louw DF, Beedle AM, Hamid J, Sutherland GR, Zamponi GW (2004) Expression of T-type calcium channel splice variants in human glioma. Glia 48:112–119

Leresche N, Hering J, Lambert RC (2004) Paradoxical potentiation of neuronal T-type Ca^{2+} current by ATP at resting membrane potential. J Neurosci 24:5592–5602

Li L, Davie JR (2010) The role of Sp 1 and Sp 3 in normal and cancer cell biology. Ann Anat 192:275–283

Li M, Zhang M, Huang L, Zhou J, Zhuang H, Taylor JT, Keyser BM, Whitehurst RM Jr (2005) T-type Ca^{2+} channels are involved in high glucose-induced rat neonatal cardiomyocyte proliferation. Pediatr Res 57:550–556

Li W, Zhang S-L, Wang N, Zhang B-B, Li M (2011) Blockade of T-type Ca^{2+} channels inhibits human ovarian cancer cell proliferation. Cancer Invest 29:339–346

Lue TF (2000) Erectile dysfunction-drug therapy. N Engl J Med 342:1802–1813

Marchetti C, Brown AM (1988) Protein kinase activator 1-oleoyl-2-acetyl-sn-glycerol inhibits two types of calcium currents in GH_3 cells. Am J Physiol 254:C206–C210

Marni F, Wang Y, Morishima M, Shimaoka T, Uchino T, Zheng M, Kaku T, Ono K (2009) 17 beta-estradiol modulates expression of low-voltage-activated Ca(V)3.2 T-type calcium channel via extracellularly regulated kinase pathway in cardiomyocytes. Endocrinology 150:879–888

Michels G, Er F, Eicks M, Herzig S, Hoppe UC (2006) Long-term and immediate effect of testosterone on single T-type calcium channel in neonatal rat cardiomyocytes. Endocrinology 147:5160–5169

Monteil A, Chemin J, Bourinet E, Mennessier G, Lory P, Nargeot J (2000) Molecular and functional properties of the human α1G subunit that forms T-type calcium channels. J Biol Chem 275:6090–6100

Morishima M, Wang Y, Akiyoshi Y, Miyamoto S, Ono K (2009) Telmisartan, an angiotensin II type 1 receptor antagonist, attenuates T-type Ca^{2+} channel expression in neonatal rat cardiomyocytes. Eur J Pharmacol 609:105–112

Nelson MT, Joksovic PM, Su P, Kang HW, Van Deusen A, Baumgart JP, David LS, Snutch TP, Barrett PQ, Lee JH, Zorumski CF, Perez-Reyes E, Todorovic SM (2007a) Molecular mechanisms of subtype-specific inhibition of neuronal T-type calcium channels by ascorbate. J Neurosci 27:12577–12583

Nelson MT, Kang HW, Vitko L, Barrett PQ, Perez-Reyes E, Lee JH, Shin HS, Todorovic SM (2007b) Reducing agents sensitize C-type nociceptors by relieving high-affinity zinc inhibition of T-type calcium channels. J Neurosci 27:8250–8260

Novara M, Baldelli P, Cavallari D, Carbelli V, Giancippoli A, Carbone E (2004) Exposure to cAMP and β-adrenergic stimulation recruits Ca_v3 T-type channels in rat chromaffin cells through Epac cAMP-receptor proteins. J Physiol 558:433–449

Pachuau J, Martin-Caraballo M (2007) Extrinsic regulation of T-type Ca^{2+} channel expression in chick nodose ganglion neurons. Dev Neurobiol 67:1915–1931

Panner A, Cribbs LL, Zainelli GM, Origitano TC, Singh S, Wuster RD (2005) Variation of T-type calcium channel protein expression affects cell division of cultured tumor cells. Cell Calcium 37:105–119

Park JY, Jeong SW, Perez-Reyes E, Lee JH (2003) Modulation of Ca(v) 3.2 T-type Ca^{2+} channels by protein kinase C. FEBS Lett 547:37–42

Park JY, Kang HW, Moon HJ, Huh SU, Jeong SW, Soldatov NM, Lee JH (2006) Activation of protein kinase C augments T-type Ca^{2+} channel activity without changing channel surface density. J Physiol 577:513–523

Pemberton KE, Hill-Eubanks LJ, Jones SV (2000) Modulation of low-threshold T-type calcium channels by the five muscarinic receptor subtypes in NIH 3 T3 cells. Pflugers Arch 440:452–461

Perez-Reyes E (2003) Molecular physiology of low-voltage-activated T-type calcium channels. Physiol Rev 83:117–161

Plueanu F, Cribbs LL (2011) Regulation and function of $Ca_v3.1$ T-type calcium channels in IGF-I-stimulated pulmonary artery smooth muscle cells. Am J Physiol Cell Physiol 300:C517–C525

Pottle J, Sun C, Gray L, Li M (2013) Exploiting MCF-7 cell's calcium dependence with interlaced therapy. J Cancer Ther 4:32–40

Rangel A, Sánchez-Armass S, Meza U (2010) Protein kinase C-mediated inhibition of recombinant T-type $Ca_V3.2$ channels by neurokinin 1 receptors. Mol Pharmacol 77:202–210

Rao F, Deng C-Y, Wu S-L, Xiao D-Z, Huang W, Deng H, Kuang S-J, Lin Q-X, Shan Z-X, Liu X-Y, Zhu J-N, Yu X-Y (2013) Mechanism of macrophage migration inhibitory factor-induced decrease of T-type Ca^{2+} channel current in atrium-derived cells. Exp Physiol 98:172–182

Rbinovitch A, Suarez-Pinzon WL, Shi Y, Morgan AR, Bleackley RC (1994) DNA fragmentation is an early event in cytokine-induced islet β-cell destruction. Diabetologia 37:733–738

Richard S, Neveu D, Carnac G, Bodin P, Travo P, Nargeöt J (1992) Differential expression of voltage-gated Ca^{2+} currents in cultivated aortic myocytes. Biochem Biophys Acta 1160:95–104

Ritchie AK (1993) Estrogen increases low voltage-activated calcium current density in GH_3 anterior pituitary cells. Endocrinology 132:1621–1629

Rossier MF, Lesouhaitier O, Perrier E, Bockhorn L, Chiappe A, Lalevée N (2003) Aldosterone regulation of T-type calcium channels. J Steroid Biochem Mol Biol 85:282–288

Schroeder JE, Fischbach PS, McClesky EW (1990) T-type calcium channels: heterogeneous expression in rat sensory neurons and selective modulation by phorbol esters. J Neurosci 10:947–951

Sen L, Smith TW (1994) T-type Ca^{2+} channels are abnormal in genetically determined cardiomyophathic hamster hearts. Curr Res 75:149–155

Shan HQ, Hammarback JA, Godwin DW (2013) Ethanol inhibition of a T-type Ca^{2+} channel through activity of protein kinase C. Alcohol Clin Exp Res 37:1333–1342

Sun-Un H, Kang H-W, Park J-Y, Lee J-H (2008) Regulation of $Ca_v3.2$ Ca^{2+} channel activity by protein tyrosine phosphorylation. J Microbiol Biotechnol 18:365–368

Suske G (1999) The Sp-family of transcription factors. Gene 238:291–300

Tao J, Hildebrand ME, Liao P, Liang MC, Tan G, Shengnan L, Snutch TP, Soong TW (2008) Activation of corticotropin-releasing factor receptor 1 selectively inhibits Cav3.2 T-type calcium channels. Mol Pharmacol 73:1596–1609

Taylor JT, Huang L, Pottle JE, Liu K, Yang Y, Zeng X, Keyser BM, Agrawal KC, Hansen JB, Li M (2008a) Selective blockage of T-type Ca^{2+} channels suppresses human breast cancer cell proliferation. Cancer Lett 267:116–124

Taylor JT, Zeng XB, Pottle JE, Lee K, Wang AR, Yi SG, Scruggs JAS, Sikka SS, Li M (2008b) Calcium signaling and T-type calcium channels in cancer cell cycling. World J Gastroenterol 14:4984–4991

Todorovic SM, Jevtovic-Todorovic V, Mennerick S, Perez-Reyes E, Zorumski CF (2001) $Ca_v3.2$ channel is a molecular substrate for inhibition of T-type calcium currents in rat sensory neurons by nitrous oxide. Mol Pharmacol 60:603–610

Toyota M, Ho C, Ohe-Toyota M, Baylin SB, Issa JP (1999) Inactivation of CACNA1G, a T-type calcium channel gene, by aberrant methylation of its 5' CpG island in human tumors. Cancer Res 59:4535–4541

Traboulsie A, Chemin J, Chevalier J-F, Nargeot J, Lory P (2007) Subunit-specific modulation of T-type calcium channels by zinc. J Physiol 578:159–171

Trimarchi T, Pachuau J, Shepherd A, Dey D, Martin-Caraballo M (2009) CNTF-evoked activation of JAK and ERK mediates the functional expression of T-type Ca^{2+} channels in chicken nodose neurons. J Neurochem 108:246–259

Tseng GN, Boyden PA (1991) Differential effects of intracellular Ca and protein kinase C on cardiac T and L Ca currents. Am J Physiol 262:H364–H379

Wang R, Karpinski E, Pang PK (1991) Two types of voltage-dependent calcium channel currents and their modulation by parathyroid hormone in neonatal rat ventricular cells. J Cardiovasc Pharmacol 17:990–998

Wang L, Bhattacharjee A, Zuo Z, Hu F, Honkanen RE, Berggren P-O, Li M (1999) A low voltage-activated Ca^{2+} current mediates cytokine-induced pancreatic β-cell death. Endocrinology 140:1200–1204

Wang Y, Morishima M, Zheng M, Uchino T, Mannen K, Tkahashi A, Nakaya Y, Komuro I, Ono K (2007) Transcription factors CsX/Nkx2.5 and GATA4 distinctly regulate expression of Ca^{2+} channels in neonatal rat heart. J Mol Cell Cardiol 42:1045–1053

Welsby PJ, Wang H, Wolfe JT, Colbran RJ, Johnson ML, Barrett PQ (2003) A mechanism for the direct regulation of T-type calcium channels by Ca^{2+}/calmodulin-dependent kinase II. J Neurosci 23:10116–10121

Wolfe JT, Wang H, Howard J, Garrison JC, Barrett PQ (2003) T-type calcium channel regulation by specific G-protein βγ subunits. Nature 424:209–213

Wuerstra I (2008) Sp 1: emerging roles beyond constitutive activation of TATA-less housekeeping genes. Biochem Biophys Res Commun 372:1–13

Yao J, Davies LA, Howard JD, Adney SK, Welsby PJ, Howell N, Robert M, Carey RM, Colbran RJ, Barrett PQ (2006) Molecular basis for the modulation of native T-type Ca^{2+} channels *in vivo* by Ca^{2+}/calmodulin-dependent protein kinase II. J Clin Invest 116:2403–2412

Zeng X, Keyser B, Li M, Sikka SC (2005) T-type (α1G) low voltage activated calcium channel interaction with nitric oxide-cyclic guanosine monophosphate pathway and regulation of calcium homeostasis in human cavernosal cells. J Sex Med 2:620–633

Zhang M, Zhuang H, Bhattacharjee A, Li M (2000a) High glucose elevated T-type calcium channel expression and basal $[Ca^{2+}]_i$ in rat islet beta cells. Biophys J 78:69A

Zhang Y, Cribbs LL, Satin J (2000b) Arachidonic acid modulation of α1H, a cloned human T-type calcium channel. Am J Physiol 278:H184–H193

Pharmacological Profiles of T-Type Calcium Channel Antagonists

3

Norio Hashimoto and Takeshi Kawazu

Abstract

The activation of the channel alters the occurrence and severity of many diseases, including hypertension, arrhythmias, pain, epilepsy, and cancer. Therefore, pharmaceutical companies have been trying to develop a selective blocker of the T-type Ca^{2+} channels. Several T-type Ca^{2+} channel blockers exist, including mibefradil, dihydropyridine derivatives, and antiepilepsy agents. This group of drugs is moderately selective for the T-type Ca^{2+} channels, with some of them also involved in untoward drug–drug interactions. T-type Ca^{2+} channel blockers under clinical development include mibefradil and Z-944, which inhibit all T-type Ca^{2+} channel subtypes, human Cav3.1, Cav3.2, and Cav3.3, with an IC_{50} value ranging from 50 to 160 nM. T-type calcium channel blockers presently in preclinical research include A-1048400, TTA-A2, TTA-P2, KST-5468, KYS05090, 2-(1-alkylpiperidin-4-yl)-*N*-[(1*R*)-1-(4-fluorophenyl)-2-methylpropyl] acetamide derivatives, *R*(−)-efonidipine, *N*-[[1-[2-(tert-butylcarbamoylamino)ethyl]-4-(hydroxymethyl)-4-piperidyl]methyl]-3,5-dichlorobenzamide, and mibefradil derivatives. The new group of drugs exhibit high potency and high selectivity for the T-type Ca^{2+} channels, thereby possessing promising potential futures.

N. Hashimoto (✉)
Pharmaceutical Division, Nissan Chemical Industries, Ltd, Tokyo, Japan
e-mail: hashimoton@nissanchem.co.jp

T. Kawazu
Pharmaceutical Research Department, Biological Research Laboratories, Nissan Chemical Industries, Ltd, Saitama, Japan

S.W. Schaffer, M. Li (eds.), *T-type Calcium Channels in Basic and Clinical Science*,
DOI 10.1007/978-3-7091-1413-1_3

3.1 Introduction

The T-type calcium channel (TCC) is a low-voltage-activated Ca^{2+} channel that plays a crucial role in the regulation of cellular excitability, cell cleavage, and muscle contraction. The activation of the channel is reportedly related to the occurrence and maintenance of some diseases, such as hypertension, arrhythmias, pain, epilepsy, and cancer. Therefore, pharmaceutical companies have been trying to develop a selective blocker of the TCC. The first launched T-type calcium channel blocker was "mibefradil," which was approved for the treatment of hypertension and angina. Unfortunately, this drug was withdrawn due to interactions with other drugs. However, some researchers have revealed that TCC blockade with mibefradil is effective in several different animal models, not only against cardiovascular diseases but also for the treatment of cancer, pain, epilepsy, etc. Therefore, the TCC is expected to be a fascinating target for a wide range of diseases. In this chapter, we will discuss the characteristics and development status of existing and newly developed TCC antagonists.

3.2 Existing T-Type Calcium Channel Blockers

3.2.1 Mibefradil

Mibefradil (the structure is shown in Fig. 3.1a), which was developed by F. Hoffmann-La Roche Ltd. (Basel, Switzerland), was launched as a drug exhibiting moderately selective TCC inhibitory activity. The drug was approved for hypertension and chronic angina pectoris in 1996 but was voluntarily withdrawn from the market in 1998 because of the potential risk for rhabdomyolysis, renal failure, and bradycardia when given concomitantly with other drugs (Paoletti et al. 2002). Prueksaritanont et al. (1999) showed that this drug is a potential mechanism-based inactivator of CYP3A4. However, as a single agent, it was well tolerated compared to other antihypertensive drugs (Kobrin et al. 1997). In addition, end-organ damage, such as myocardial hypertrophy, was prevented by mibefradil administration (Schmitt et al. 1995). These results suggest that a TCC blocker can be a safe, more effective treatment of hypertension. Even after withdrawal, mibefradil has been used as an effective tool for TCC researchers. The results from the studies using mibefradil suggest that TCC plays a pathophysiological role

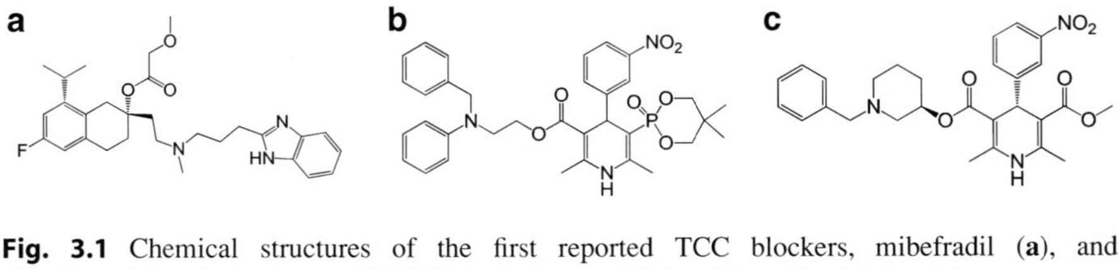

Fig. 3.1 Chemical structures of the first reported TCC blockers, mibefradil (**a**), and dihydropyridine derivatives with TCC blockade: efonidipine (**b**) and benidipine (**c**)

in a wide range of diseases other than cardiovascular diseases, including cancer, pain, and epilepsy.

3.2.2 Dihydropyridine Derivatives

Some dihydropyridine derivatives reportedly block not only L-type but also T-type calcium channels. These blockers, including efonidipine (developed by Nissan Chemical Industries, LTD., Tokyo, Japan, Fig. 3.1b) and benidipine (developed by Kyowa-Hakko CO, Machida, Japan, Fig. 3.1c), have different pharmacological characteristics from standard calcium channel antagonists.

Masumiya et al. (1998) reported that efonidipine at high concentrations (~1 μM) inhibits TCC current in ventricular myocytes and produces an increase in cycle length accompanied by prolongation of phase 4 depolarization. Efonidipine also inhibits heart rate elevation with favorable effects on the autonomic nervous system in mild to moderate essential hypertensive patients (Harada et al. 2003). These results suggest that efonidipine may exert a protective influence on the heart. This drug also exhibits renal protective activity. Eight-week treatment of efonidipine markedly prevents the increase in proteinuria in partially nephrectomized spontaneously hypertensive rats, whereas nifedipine does not affect proteinuria in a similar setting (Fujiwara et al. 1998). This beneficial effect is possibly mediated through afferent arteriolar dilation and a decrease in glomerular capillary pressure (Honda et al. 2001; Nakamura et al. 2002). In addition, the dihydropyridine derivatives exhibiting TCC blockade inhibit aldosterone synthesis and production in a human aldosterone cell line (Imagawa et al. 2006; Isaka et al. 2009) and decrease plasma aldosterone levels in humans (Okayama et al. 2006). These effects support the prominence of dual L- and T-type calcium channel blockers relative to L-type calcium channel selective blockers in protecting against cardiovascular and renal dysfunction.

3.2.3 Antiepilepsy Agents

Some of the conventional antiepileptic drugs, such as ethosuximide, phenytoin, valproate sodium, and zonisamide (Fig. 3.2a–d), reportedly inhibit TCC (Suzuki et al. 1992; Todorovic and Lingle 1998). The TCC appears to play a crucial role in the regulation of neuronal firing and low-amplitude voltage oscillations. The current has also been implicated in seizure susceptibility and initiation (Huguenard and Prince 1994; Tsakiridou et al. 1995). More recently, eslicarbazepine acetate (Fig. 3.2e) was accepted by FDA for use as once-daily adjunctive therapy in the treatment of partial-onset epileptic seizure in patients 18 years and older. This compound reportedly blocks both sodium channels and TCC (Brady et al. 2011; Hebeisen et al. 2011). The TCC blocking effects of these agents are not particularly potent even when applied at higher concentrations. Moreover, these drugs have broad pharmacological activity. The clinical contribution of TCC blockade in the

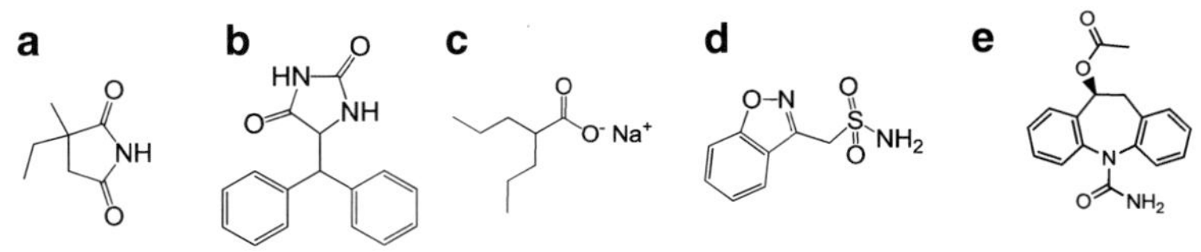

Fig. 3.2 Structures of antiepilepsy drugs with TCC inhibition, ethosuximide (**a**), phenytoin (**b**), valproate sodium (**c**), zonisamide (**d**), and eslicarbazepine (**e**)

treatment of seizures should be investigated using more specific and potent TCC antagonists.

3.3 T-Type Calcium Channel Blockers Under Clinical Development

3.3.1 Mibefradil (Redevelopment)

T-type calcium channels play a direct role in regulating calcium concentration, especially in non-excitable tissues, including some cancerous cells (Lory et al. 2006). Therefore, TCC blockade may lead to aberrant cell growth and tumor progression. Recently, sequential administration of a TCC blocker with the aim of synchronizing cells at the G1/S checkpoint of the cell cycle before the administration of chemotherapy (Interlaced Therapy™) has been reported by Tau Therapeutics LLC (Charlottesville, USA). Using the interlaced approach, the combination of mibefradil and temozolomide was found to be more effective than either single agent against brain tumor (glioblastoma multiforme) cells from D-54 MG and 43 MG xenograft mice. The increased benefits of combined therapy in the treatment of tumors of D-54 MG and 43 MG mice in comparison to temozolomide therapy resulted in life extension of an additional 150 days and 8.9 days, respectively (Keir et al. 2013). Tau Therapeutics has completed a phase I study of a repurposed drug, mibefradil, in normal healthy volunteers. A phase Ib study in patients with recurrent high-grade glioma is being conducted in conjunction with the National Cancer Institute. Tau Therapeutics has also been developing its proprietary lead compound, TLL-1177, which at the present time is in the preclinical testing stage.

3.3.2 Z-944

Z-944 (Fig. 3.3a) is a state-dependent TCC blocker developed by Zalicus Pharmaceuticals Ltd. (Vancouver, Canada). Under conditions that produce 30 % channel inactivation, Z-944 inhibits all TCC subtypes, human Cav3.1, Cav3.2, and Cav3.3, with IC_{50} values of 50–160 nM (Tringham et al. 2012). In the closed state of the channel, the compound blocks human Cav3.1, Cav3.2, and Cav3.3 with IC_{50}

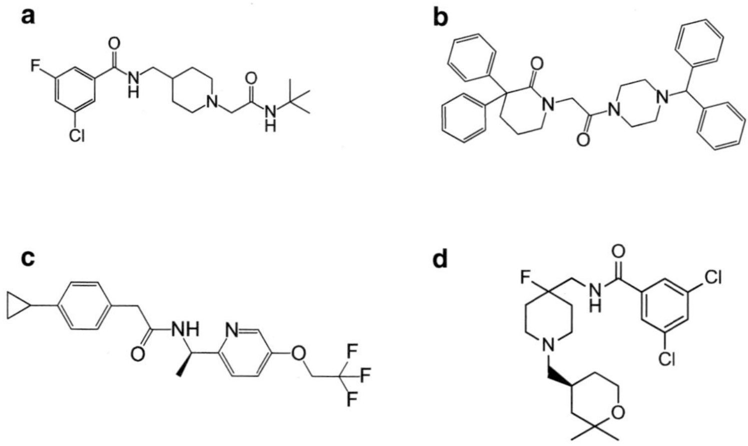

Fig. 3.3 Chemical structures of novel TCC blockers; Z-944 (**a**), A-1048400 (**b**), TTA-A2 (**c**), and TTA-P2 (**d**)

values ranging from 130 to 540 nM, which are 2.5- to 4-fold lower than those for the inactivated state. These effects of Z-944 on inactivated TCC currents are 70 times more potent than those seen for channel blockade ($IC_{50} = 11$ μM). Z-944 significantly reduces time spent in seizure and at the highest dose (30 mg/kg), almost completely suppresses the onset of seizures e (85–90 %) in the Genetic Absent Epilepsy Rats from Strasbourg (GAERS) model. Based on the duration and the cycle frequency of spike-and-wave discharges in the GAERS model, Z-944 acts via a different mechanism in the thalamocortical circuitry than do the two first-line clinical drugs, ethosuximide and valproate.

Z-944 has been selected for progression into phase I human studies to assess its safety, exposure, and efficacy. According to the press release from Zalicus, the company plans to conduct a phase Ib study using a state-of-the-art experimental clinical model utilizing laser-evoked potential to provide an efficient and objective assessment of the activity of Z-944 against induced neuropathic pain.

3.4 T-Type Calcium Channel Blockers Under Preclinical Research

3.4.1 A-1048400

A-1048400 (Fig. 3.3b) is a diphenyl lactam variant developed by Abbott Laboratories (Abbott Park, USA) that blocks T-type and N-type calcium channel currents in whole-cell patch clamp recordings of HEK293 cells overexpressing human Cav3.2 and Cav2.2, respectively (Scott et al. 2012). In the closed state, A-1048400 blocks TCC current with an $IC_{50} = 2.6$ μM and N-type calcium channel (NCC) current with an $IC_{50} = 1.4$ μM. In the inactivated state, this compound blocks TCC current with an $IC_{50} = 2.6$ μM and N-type calcium channel (NCC) current with $IC_{50} = 1.4$ μM. This compound also blocks native TCC and NCC currents in rat dorsal root ganglion neurons ($IC_{50} = 3.0$ μM and 1.4 μM, respectively). On the other hand, this compound is less effective in altering L-type calcium channel (LCC) current ($EC_{50} = 28$ μM).

A-1048400 is shown to be effective in several kinds of pain models. Oral administration of A-1048400 (28 mg/kg) reverses tactical allodynia in a model of capsaicin-induced secondary hypersensitivity (Scott et al. 2012). A-1048400 also attenuates mechanical allodynia with an $ED_{50} = 38$ mg/kg, p.o. in the complete Freunds' adjuvant-induced inflammation pain model. Furthermore, this compound attenuates mechanical allodynia caused by spinal nerve ligation and chronic constriction injury of the sciatic nerve ($ED_{50} = 15$ mg/kg, p.o. and 53 mg/kg, p.o., respectively). However, A-108400 did not show significant untoward actions against hemodynamic parameters, such as mean arterial pressure, heart rate, and left ventricular contractility. In addition, the compound did not decrease balance or motor performance at doses up to 300 mg/kg, p.o. Therefore, A-1048400 appears to exhibit robust antinociception with minimum effects on cardiovascular and psychomotor functions.

3.4.2 TTA-A2, TTA-P2

TTA-A2 (Fig. 3.3c) is a state-dependent TCC blocker developed by Merck Research Laboratories (West Point, USA, Uebele et al. 2009). In whole-cell patch recordings of human Cav3.2 expressed in HEK293 cells, TTA-A2 inhibited TCC currents with $EC_{50} = 8.99$ nM and 22.6 μM at holding potentials of −75 mV and −110 mV, respectively (Francois et al. 2013). The compound showed a 3.3-fold selectivity for Cav3.2 compared to Cav3.1 ($EC_{50} = 30$ nM and 100.6 nM for Cav3.2 and Cav3.1 at a holding potential of −75 mV, respectively). TTA-A2 also blocked native low-voltage-activated Ca^{2+} current with an $EC_{50} = 22.7$ nM at a holding potential of −75 V in medium-sized dissociated mouse dorsal root ganglion neurons, but did not block high-voltage-activated Ca^{2+} current. In addition, TTA-A2 increased action potential thresholds of dorsal root ganglion neurons from wild-type mice that express TCC, but did not change the action potential thresholds of neurons from Cav3.2 knockout mice.

TTA-A2 increases acute thermal and mechanical sensitivity in wild-type mice, but not in Cav3.2 knockout mice. TTA-A2 also attenuates hypersensitivity in an irritable bowel syndrome model induced by butyrate treatment, demonstrating its therapeutic potential for the treatment of pathological pain (Francois et al. 2013). In addition, TTA-A2 reportedly inhibits native TCC currents in brain slice recordings from the dorsal lateral geniculate nucleus and the subthalamic nucleus, suggesting that this compound may have a sleep-inducing effect (Kraus et al. 2010). Finally, TTA-A2 suppresses the state of wakefulness and promotes slow-wave sleep in wild-type mice but not in mice lacking both Cav3.1 and Cav3.3.

Another TCC blocker reported by Merck is TTA-P2 (Fig. 3.3d), a 4-aminomethyl-4-fluoropiperidine derivative (Choe et al. 2011). This compound blocks TCC at a holding potential of −90 mV in dorsal root ganglion cells acutely dissociated from rats. On the other hand, TTA-P2 has little effect on high-voltage-activated Ca^{2+} currents, human recombinant Cav2.3 current, and voltage-gated sodium currents. TTA-P2 (7.5 mg/kg, i.p.) also significantly reduces pain responses in the first and second phases of the formalin test. Moreover, TTA-P2 (10 mg/kg, i.p.) completely reverses thermal hyperalgesia in streptozotocin-induced diabetic animals. On the other hand, this compound (7.5 mg/kg, i.p.) does not change sensorimotor activity.

3.4.3 KST-5468

KST-5468 (Fig. 3.4a) is a TCC blocker developed by Seoul National University (Seoul, Republic of Korea) to exert antinociceptive effects in several animal pain models (Lee et al. 2010). In the hot plate test, peritoneal administration of KST-5468 (10 mg/kg) significantly increases the withdrawal latency of noxious thermal stimuli. This compound also significantly decreases pain responses in the first and second phases of the formalin test. Moreover, KST-5468 significantly attenuates mechanical hypersensitivity in two types of neuropathic pain models induced by chronic constriction injury or spared nerve injury. Using immunohistochemistry, expression of two well-known pain-related molecular markers, c-Fos and calcitonin gene-related peptide (CGRP), and phosphorylated extracellular signal-related kinase (p-ERK) were found to be decreased in the laminae I–II layers

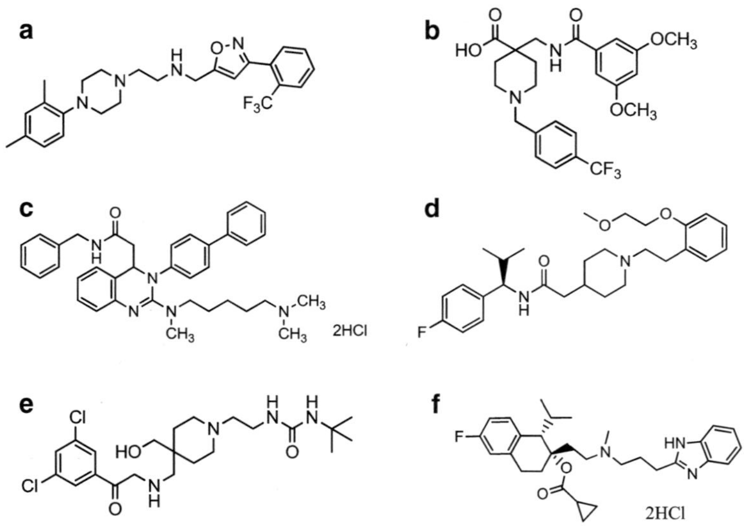

Fig. 3.4 Chemical structures of novel TCC blockers; KST-5468 (**a**), compound 31a (**b**), KYS05090 (**c**), compound 4f (**d**), compound 21 (**e**), and NNC 55-0396 (**f**)

of the ipsilateral L4–L5 spinal dorsal horn of KST-5468-treated mice. Taken together, KST-5468 may be an effective antinociceptive agent for treatment of inflammatory and neuropathic pain.

The research team from Seoul National University and the Korean Institute of Science and Technology (Seoul, Republic of Korea) introduced several types of TCC blockers. One of their lead compounds (compound 31a, Fig. 3.4b) is a 4-piperidinecarboxylate and 4-piperidine cyanide derivative that inhibits Cav3.1 current with an $IC_{50} = 7.31$ μM (Woo et al. 2011). Oral administration of this compound (100 mg/kg) inhibits mechanical allodynia in the spinal nerve ligation model of neuropathic pain.

3.4.4 KYS05090

Kyung Hee University (Seoul, Republic of Korea) reported that their 3,4-dihydroquinazoline derivative, KYS05090 (Fig. 3.4c), is a selective TCC blocker that decreases tumor size 49 % compared with vehicle when administered intravenously (2 mg/kg) to A549 xenograft mice (Jung et al. 2010). In addition, administration of this drug (1 or 5 mg/kg, five times a week; p.o.) to A549 xenograft mice exhibits good oral bioavailability and more potent antitumor activity than paclitaxel (10 mg/kg, twice a week; i.p.) (Kang et al. 2012). The university has optimized an efficient scale-up procedure for large production of this compound and has prepared ca. 100 g of KYS05090 for further preclinical trials.

3.4.4.1 2-(1-Alkylpiperidin-4-yl)-*N*-[(1*R*)-1-(4-Fluorophenyl)-2-Methylpropyl] Acetamide Derivatives

Astellas Pharma Inc. (Tokyo, Japan) have synthesized and evaluated a series of 2-(1-alkylpiperidin-4-yl)-*N*-[(1*R*)-1-(4-fluorophenyl)-2-methylpropyl] acetamide derivatives that exhibit TCC selective blockade (Watanuki et al. 2012). One of the derivatives, *N*-[(1*R*)-1-(4-fluorophenyl)-2-methylpropyl]-2-(1-{2-[2-(2-methoxyethoxy) phenyl] ethyl}-piperidin-4-yl) acetamide (compound 4f, Fig. 3.4d), exhibits equipotent TCC current inhibition ($IC_{50} = 0.18$ μM) to mibefradil ($IC_{50} = 0.12$ μM). However, this compound shows poor inhibitory activity against L-type calcium current ($IC_{50} = 38$ μM), with the IC_{50} being tenfold less potent than that of mibefradil ($IC_{50} = 3.2$ μM). When administrated orally (3 or 10 mg/kg) to spontaneously hypertensive rats, the compound exerts an antihypertensive effect (10–40 % decrease in mean blood pressure) without associated reflex tachycardia. In contrast, mibefradil exhibits a slight increase in heart rate (12 %) due to insufficient TCC selectivity. Thus, TCC selective blockers appear to be novel antihypertensive drugs that lack adverse effects, such as tachycardia induced by traditional L-type calcium channel blockers.

3.4.5 *R*(−)-Efonidipine

A dihydropyridine derivative, efonidipine, is a racemic compound, with the *R*(−)-enantiomer (*R*(−)-efonidipine) exhibiting a TCC selective profile but weak, poor L-type calcium channel inhibitory activity. Furukawa et al. (2004) reported that the IC_{50} for *R*(−)-efonidipine blockade of the L-type calcium channels (>1,000 μM) was more than 100-fold larger than that of the T-type calcium channel (~1 μM) at a holding potential of −100 mV *Xenopus* oocytes and BHK cells expressing T-type calcium channels. In guinea pig ventricular myocytes, *R*(−)-efonidipine exhibits concentration-dependent blockade of TCC current with 85 % inhibition seen at 1 μM, but no effect on L-type calcium current and Ca^{2+} transients of cardiomyocytes and contractile force of papillary muscle (Tanaka et al. 2004). In partially nephrectomized spontaneously hypertensive rats, this compound significantly reduced proteinuria, blunted the increase in tubulointerstitial fibrosis, and suppressed renal expression of alpha-smooth muscle actin and vinectin without affecting blood pressure (Sugano et al. 2008). More recently, Hu et al. (2012) reported that *R*(−)-efonidipine attenuates angiotensin II-mediated reductions in renal medullary blood flow in rats, an action that may protect against ischemic renal injury in the renal medullary region.

3.4.5.1 *N*-[[1-[2-(Tert-Butylcarbamoylamino)Ethyl]-4-(Hydroxymethyl)-4-Piperidyl]Methyl]-3,5-Dichloro-Benzamide

Recently, AstraZeneca R&D CVGI iMed (Molndal, Sweden) introduced several novel, selective, and peripherally restricted TCC blockers (Giordanetto et al. 2013). They designed the compounds by focusing on minimizing exposure to the central nerve system and improving their pharmacokinetic properties while maintaining adequate potency and selectivity. The lead compound for in vivo profiling (compound 21, Fig. 3.4e) blocks human Cav3.2 current with an $IC_{50} = 0.6–1.2$ μM. The compound showed highly selective TCC blockade in comparison to other ion channels (cardiac relevant ion channels (hERG, Nav1.5, HCN4, IKs, Kv4.3) $IC_{50} > 33$ μM and Cav1.2 (LCC) $IC_{50} > 31.6$ μM). Compound 21 reduces heart rate during the infusion in a concentration-dependent fashion (~20 % decrease at the plasma concentration of 0.57 μM). Giordanetto et al (2013) concluded that compound 21 could be a good candidate for a proof-of-concept study in dogs to verify the cardioprotective role of Cav3.2 blockade in the treatment of atrial fibrillation.

3.4.6 Mibefradil Derivatives

The most important adverse property leading to mibefradil withdrawal has been drug-drug interactions. Thus, some companies tried to develop mibefradil analogs showing less cytochrome P450 inhibition. NNC 55-0396 (Fig. 3.4f) is a mibefradil derivative that retains potent in vitro TCC antagonist efficacy with less cytochrome P450 inhibition than mibefradil. This compound blocks recombinant TCC (Cav3.1) in HEK293 cells ($IC_{50} = \sim 7\ \mu M$), whereas 100 μM NNC 55-0396 has no detectable effect on high-voltage-activated channels (Huang et al. 2004). Li et al. (2011) reported the compound slowed ovarian cancer development in nude mice. Concert Pharmaceuticals Inc. (Lexington, USA) reported that the deuterium-enriched analogs of mibefradil, C-10296, and C-10302 exhibit a reduced propensity for CYP3A4-mediated drug-drug interactions according to in vitro assays using human liver microsomes. Furthermore, these deuterated analogs inhibited CYP3A4-mediated atorvastatin hydroxylation less than mibefradil.

Conclusions

Preclinical studies suggest that T-type calcium channels may play a critical role in the pathogenesis of certain diseases, including pain, cancer, and epilepsy. However, selective TCC blockers have not been available for clinical applications. Many pharmaceutical companies and bioventures are trying to develop novel TCC compounds with high selectivity. Some companies have undertaken some human clinical trials. In the near future, the utility of TCC blockers should be elucidated.

References

Brady K, Hebeisen S, Konrad D, Soares-da-Silva P (2011) The effects of eslicarbazepine, *R*-licarbazepine, oxcarbazepine and carbamazepine on ion transmission $Ca_v3.2$ channels. Epilepsia 52(Suppl 6):260

Choe W, Messinger RB, Leach E, Eckle VS, Obradovic A, Salajegheh R, Jevtovic-Todorovic V, Todorovic SM (2011) TTA-P2 is a potent and selective blocker of T-type calcium channels in rat sensory neurons and a novel antinociceptive agent. Mol Pharmacol 80(5):900–10

Francois A, Kerckhove N, Meleine M, Alloui A, Barrere C, Gelot A, Uebele VN, Renger JJ, Eschalier A, Ardid D, Bourinet E (2013) State-dependent properties of a new T-type calcium channel blocker enhance Ca(V)3.2 selectivity and support analgesic effects. Pain 154:283–93

Fujiwara K, Kanno Y, Hayashi K, Takenaka T, Saruta T (1998) Renal protective effects of efonidipine in partially nephrectomized spontaneously hypertensive rats. Clin Exp Hypertens 20:295–312

Furukawa T, Miura R, Honda M, Kamiya N, Mori Y, Takeshita S, Isshiki T, Nukada T (2004) Identification of R(-)-isomer of efonidipine as a selective blocker of T-type Ca2+ channels. Br J Pharmacol 143:1050–7

Giordanetto F, Wållberg A, Knerr L, Selmi N, Ullah V, Thorstensson F, Lindelöf Å, Karlsson S, Nikitidis G, Llinas A, Wang QD, Lindqvist A, Högberg G, Lindhardt E, Åstrand A, Duker G (2013) Discovery of N-[[1-[2-(tert-butylcarbamoylamino)ethyl]-4-(hydroxymethyl)-4-piperidyl]methyl]-3,5-dichloro-benzamide as a selective T-type calcium channel (Cav3.2) inhibitor. Bioorg Med Chem Lett 23:119–24

Harada K, Nomura M, Nishikado A, Uehara K, Nakaya Y, Ito S (2003) Clinical efficacy of efonidipine hydrochloride, a T-type calcium channel inhibitor, on sympathetic activities. Circ J 67:139–45

Hebeisen S, Brady K, Konrad D, Soares-da-Silva P (2011) Inhibitory effects of eslicarbazepine acetate and its metabolites against neuronal voltage-gated sodium channels. Epilepsia 52 (Suppl 6):257–8

Honda M, Hayashi K, Matsuda H, Kubota E, Tokuyama H, Okubo K, Takamatsu I, Ozawa Y, Saruta T (2001) Divergent renal vasodilator action of L- and T-type calcium antagonists in vivo. J Hypertens 19:2031–7

Hu C, Mori T, Lu Y, Guo Q, Sun Y, Yoneki Y, Ohsaki Y, Nakamichi T, Oba I, Sato E, Ogawa S, Dickinson BC, Chang CJ, Miyata T, Sato H, Ito S (2012) Role of specific T-type calcium channel blocker R(-) efonidipine in the regulation of renal medullary circulation. J Hypertens 30:1620–31

Huang L, Keyser BM, Tagmose TM, Hansen JB, Taylor JT, Zhuang H, Zhang M, Ragsdale DS, Li M (2004) NNC 55-0396 [(1S,2S)-2-(2-(N-[(3-benzimidazol-2-yl)propyl]-N-methylamino) ethyl)-6-fluoro-1,2,3,4-tetrahydro-1-isopropyl-2-naphtyl cyclopropanecarboxylate dihydrochloride]: a new selective inhibitor of T-type calcium channels. J Pharmacol Exp Ther 309:193–9

Huguenard JR, Prince DA (1994) Intrathalamic rhythmicity studied in vitro: nominal T-current modulation causes robust antioscillatory effects. J Neurosci 14:5485–502

Imagawa K, Okayama S, Takaoka M, Kawata H, Naya N, Nakajima T, Horii M, Uemura S, Saito Y (2006) Inhibitory effect of efonidipine on aldosterone synthesis and secretion in human adenocarcinoma (H295R) cells. J Cardiovasc Pharmacol 47:133–8

Isaka T, Ikeda K, Takada Y, Inada Y, Tojo K, Tajima N (2009) Azelnidipine inhibits aldosterone synthesis and secretion in human adrenocortical cell line NCI-H295R. Eur J Pharmacol 605:49–52

Jung SY, Lee SH, Kang HB, Park HA, Chang SK, Kim J, Choo DJ, Oh CR, Kim YD, Seo JH, Lee KT, Lee JY (2010) Antitumor activity of 3,4-dihydroquinazoline dihydrochloride in A 549 xenograft in mice. Bioorg Med Chem Lett 20:6633–6636

Kang HB, Rim HK, Park JY, Choi HW, Choi DL, Seo JH, Chung KS, Huh G, Kim J, Choo DJ, Lee KT, Lee JY (2012) In vivo evaluation of oral anti-tumoral effect of 3,4-dihydroquinazoline derivative on solid tumor. Bioorg Med Chem Lett 22:1198–201

Keir ST, Friedman HS, Reardon DA, Bigner DD, Gray LA (2013) Mibefradil, a novel therapy for glioblastoma multiforme: cell cycle synchronization and interlaced therapy in a murine model. J Neurooncol 111(2):97–102

Kobrin I, Charlon V, Lindberg E, Pordy R (1997) Safety of mibefradil, a new once-a-day, selective T-type calcium channel antagonist. Am J Cardiol 80(4B):40C–46C

Kraus RL, Li Y, Gregan Y, Gotter AL, Uebele VN, Fox SV, Doran SM, Barrow JC, Yang ZQ, Reger TS, Koblan KS, Renger JJ (2010) In vitro characterization of T-type calcium channel antagonist TTA-A2 and in vivo effects on arousal in mice. J Pharmacol Exp Ther 335:409–17

Lee MJ, Shin TJ, Lee JE, Choo H, Koh HY, Chung HJ, Pae AN, Lee SC, Kim HJ (2010) KST5468, a new T-type calcium channel antagonist, has an antinociceptive effect on inflammatory and neuropathic pain models. Pharmacol Biochem Behav 97:198–204

Li W, Zhang SL, Wang N, Zhang BB, Li M (2011) Blockade of T-type Ca^{2+} channels inhibits human ovarian cancer cell proliferation. Cancer Invest 29:339–46

Lory P, Bidaud I, Chemin J (2006) T-type calcium channels in differentiation and proliferation. Cell Calcium 40:135–46

Masumiya H, Shijuku T, Tanaka H, Shigenobu K (1998) Inhibition of myocardial L- and T-type Ca2+ currents by efonidipine: possible mechanism for its chronotropic effect. Eur J Pharmacol 349:351–7

Nakamura M, Notoya M, Kohda Y, Yamashita J, Takashita Y, Gemba M (2002) Effects of efonidipine hydrochloride on renal arteriolar diameters in spontaneous hypertensive rats. Hypertens Res 25:751–755

Okayama S, Imagawa K, Naya N, Iwama H, Somekawa S, Kawata H, Horii M, Nakajima T, Uemura S, Saito Y (2006) Blocking T-type Ca^{2+} channels with efonidipine decreased plasma aldosterone concentration in healthy volunteers. Hypertens Res 29:493–7

Paoletti R, Corsini A, Bellosta S (2002) Pharmacological interactions of statins. Atheroscler Suppl 3:35–40

Prueksaritanont T, Ma B, Tang C, Meng Y, Assang C, Lu P, Reider PJ, Lin JH, Baillie TA (1999) Metabolic interactions between mibefradil and HMG-CoA reductase inhibitors: an in vitro investigation with human liver preparations. Br J Clin Pharmacol 47:291–8

Schmitt R, Clozel JP, Iberg N, Bühler FR (1995) Mibefradil prevents neointima formation after vascular injury in rats. Possible role of the blockade of the T-type voltage-operated calcium channel. Arterioscler Thromb Vasc Biol 15:1161–5

Scott VE, Vortherms TA, Niforatos W, Swensen AM, Neelands T, Milicic I, Banfor PN, King A, Zhong C, Simler G, Zhan C, Bratcher N, Boyce-Rustay JM, Zhu CZ, Bhatia P, Doherty G, Mack H, Stewart AO, Jarvis MF (2012) A-1048400 is a novel, orally active, state-dependent neuronal calcium channel blocker that produces dose-dependent antinociception without altering hemodynamic function in rats. Biochem Pharmacol 83:406–18

Sugano N, Wakino S, Kanda T, Tatematsu S, Homma K, Yoshioka K, Hasegawa K, Hara Y, Suetsugu Y, Yoshizawa T, Hara Y, Utsunomiya Y, Tokudome G, Hosoya T, Saruta T, Hayashi K (2008) T-type calcium channel blockade as a therapeutic strategy against renal injury in rats with subtotal nephrectomy. Kidney Int 73:826–34

Suzuki S, Kawakami K, Nishimura S, Watanabe Y, Yagi K, Seino M, Miyamoto K (1992) Zonisamide blocks T-type calcium channel in cultured neurons of rat cerebral cortex. Epilepsy Res 12:21–7

Tanaka H, Komikado C, Shimada H, Takeda K, Namekata I, Kawanishi T, Shigenobu K (2004) The R(-)-enantiomer of efonidipine blocks T-type but not L-type calcium current in guinea pig ventricular myocardium. J Pharmacol Sci 96:499–501

Todorovic SM, Lingle CJ (1998) Pharmacological properties of T-type Ca^{2+} current in adult rat sensory neurons: effects of anticonvulsant and anesthetic agents. J Neurophysiol 79:240–52

Tringham E, Powell KL, Cain SM, Kuplast K, Mezeyova J, Weerapura M, Eduljee C, Jiang X, Smith P, Morrison JL, Jones NC, Braine E, Rind G, Fee-Maki M, Parker D, Pajouhesh H, Parmar M, O'Brien TJ, Snutch TP (2012) T-type calcium channel blockers that attenuate thalamic burst firing and suppress absence seizures. Sci Transl Med 4(121):121ra19

Tsakiridou E, Bertollini L, de Curtis M, Avanzini G, Pape HC (1995) Selective increase in T-type calcium conductance of reticular thalamic neurons in a rat model of absence epilepsy. J Neurosci 15:3110–7

Uebele VN, Nuss CE, Fox SV, Garson SL, Cristescu R, Doran SM, Kraus RL, Santarelli VP, Li Y, Barrow JC, Yang ZQ, Schlegel KA, Rittle KE, Reger TS, Bednar RA, Lemaire W, Mullen FA, Ballard JE, Tang C, Dai G, McManus OB, Koblan KS, Renger JJ (2009) Positive allosteric interaction of structurally diverse T-type calcium channel antagonists. Cell Biochem Biophys 55:81–93

Watanuki S, Matsuura K, Tomura Y, Okada M, Okazaki T, Ohta M, Tsukamoto S (2012) Synthesis and pharmacological evaluation of 2-(1-alkylpiperidin-4-yl)-n-[(1r)-1-(4-fluorophenyl)-2-methylpropyl]acetamide Derivatives as Novel Antihypertensive Agents. Chem Pharm Bull (Tokyo) 60:223–34

Woo HM, Lee YS, Roh EJ, Seo SH, Song CM, Chung HJ, Pae AN, Shin KJ (2011) Synthesis and biological evaluation of 4-piperidinecarboxylate and 4-piperidinecyanide derivatives for T-type calcium channel blockers. Bioorg Med Chem Lett 21:5910–5

Part 2

Role of T-Type Ca^{2+} Channels in Disease

4 Role of T-Type Calcium Channels in Neuroendocrine Differentiation

Marine Warnier, Florian Gackière, Morad Roudbaraki, and Pascal Mariot

Abstract

Neuroendocrine cells release their secretory products into the extracellular environment via a calcium-dependent pathway. These particular cells share common morphological and molecular features, such as the expression of specific biomarkers, neurite outgrowth and dense-core secretory granules. In order to elucidate the signalling pathways leading from undifferentiated to differentiated neuroendocrine cells, the role of voltage-dependent calcium channels and central actors in excitation–secretion coupling has been comprehensively investigated. T-type calcium channels, comprising of three different molecular isoforms, appear to be one of the important calcium channel families involved in the neuroendocrine differentiation process. They also may participate in the development of neuroendocrine tumours.

4.1 Introduction

Neuroendocrine (NE) cells produce and release (neuro)hormones, neuropeptides or monoamines in the extracellular milieu using a regulated pathway in response to a specific stimulus. The NE system includes specific organs, where cells are organized into secreting tissues (adrenals or pituitary, for instance), and a diffuse system, where endocrine cells are scattered in an apparently disorganized pattern among non-secretory cells. This diffuse NE system (DNES) was originally identified in organs, such as the lung and the gut. NE cells include both neuron-

M. Warnier (✉) • F. Gackière • M. Roudbaraki • P. Mariot
Laboratoire de Physiologie Cellulaire, Inserm U1003, Université des Sciences et Technologies de Lille 1, 59650 Villeneuve d'Ascq, France

Laboratory of Excellence, Ion Channels Science and Therapeutics, Université des Sciences et Technologies de Lille 1, 59650 Villeneuve d'Ascq, France
e-mail: warnier.marine@outlook.fr

S.W. Schaffer, M. Li (eds.), *T-type Calcium Channels in Basic and Clinical Science*,
DOI 10.1007/978-3-7091-1413-1_4

like cells and endocrine cells that are all characterized by common morphological and molecular features, like the expression of specific markers such as neuropeptides, granins, neuritic extension or dense-core secretory granules. During their development, NE cells follow a differentiation pathway which, if faulty, may participate in the development of NE tumours (NET). It is therefore of particular importance to decipher the mechanisms involved in NE differentiation that are required to generate neurohormone-releasing cells.

4.2 Features of NE Cells and NE Differentiation

From a historical point of view, the concept of NE cells derives from the seminal works of Heidenhain, Kulchitsky, Masson and Hamperl in the late nineteenth century and the early twentieth century. Their studies demonstrated, from the ability of cells to be stained by silver salts (argentaffinity or argyrophily), the existence of dispersed hormone-secreting cells in non-endocrine tissues such as the intestine (for a review, see DeLellis 2001). Using histological techniques, Feyrter and Froelich (Feyrter 1938; Froelich 1949) identified the existence of many clear cells (Hellen Zellen) throughout the body and noticeably in the gut, leading to the concept of DNES. In 1944, miscellaneous sites were identified in the prostate, where they are dispersed in the normal epithelium, representing only about 0.5–2 % of the total cell population (Pretl 1944). Anton Pearse then defined NE cells as cells sharing the common ability to uptake and decarboxylate amine precursors, to convert these precursors into biogenic amines and neurotransmitters. From this property, Pearse (1968) coined the term APUD system (amine precursor uptake and decarboxylation). NE cells were all initially thought to originate only from neuroectoderm (neural crest) and were thus also termed paraneurons (Fujita et al. 1980). However, subsequent studies showed that some of them did not derive from the neural crest. These include pancreatic endocrine cells (Le Douarin 1988) or gastrointestinal enterochromaffin cells (Andrew 1974), which originate from endodermal tissue.

NE cells are usually characterized by various functional, morphological and molecular features [for a review, see Day and Salzet (2002)]. An NE cell is characterized in electron microscopy by the presence of electrodense secretory granules containing peptidergic and aminergic neurohormones. NE cells possess the capability to rapidly release their neurohormones or neurotransmitters in response to specific stimuli, such as other hormones. In this context, the acquisition of cell excitability, characterized by action potentials and voltage-dependent ion channels, is an important step towards the acquisition of a regulated secretory pathway. Morphologically, the acquisition of an NE phenotype is usually accompanied by the lengthening of neuritic extensions. NE cells therefore share many features with neurons, but do not have specialized nerve terminals or axons. From a molecular point of view, NE cells express specific proteins that may also be expressed by neurons or even by non-neuronal cells. Among these, NE markers are the members of the granin family, chromogranins A and B (CgA and CgB) and secretogranins II, III and VII. These acidic soluble proteins are stored with neuropeptides and neurotransmitters in dense-core vesicles. Other markers of NE differentiation are the enzymes prohormone convertases PC1/3 and PC2, which process various protein precursors, including granins, into functional hormones (Seidah 2011). Like granins and neuropeptides, prohormone convertases are stored in dense-core vesicles where they exert their proteolytic activity. These enzymes have therefore been considered as good markers of the NE phenotype or differentiation (Bergeron et al. 2000). These common markers, along with others (neuron-specific enolase (NSE), synaptophysin, S100), are used in pathology to identify NE tumours.

4.3 Activation of NE Differentiation

4.3.1 Cell Models

NE differentiation leads to cell phenotypes sharing characteristic features of both developing neurons (neurite extension, neuronal markers, synaptic-like vesicles) and endocrine cells (secretory granules, neurohormones). Therefore, NE differentiation has been investigated using various cell models.

One of the most studied cell models for NE differentiation is the PC12 cell line, which is a valuable model for neurosecretion studies (Westerink and Ewing 2008). Like normal chromaffin cells, PC12 cells synthesize and release monoamines such as dopamine and noradrenaline, in a calcium-regulated manner. Upon stimulation for several days with nerve growth factor (NGF), PC12 cells differentiate into neuronal-like cells (Greene and Tischler 1976). Morphological differentiation characterized by neuritic extension is associated with reduced or halted proliferation and with increased tyrosine hydroxylase activity. Pituitary adenylate cyclase-activating polypeptide (PACAP) also leads to NE differentiation characterized by similar features (for a review, see Ravni et al. (2006)). Furthermore, NE differentiation can be induced in these cells by dexamethasone, a potent synthetic glucocorticoid, which induces the overexpression of tyrosine hydroxylase, leading to a chromaffin phenotype, potentiating calcium-dependent exocytosis through an increase in vesicle content and calcium channel coupling to release sites (Elhamdani et al. 2000).

The LNCaP prostate cancer cell line is also a model of choice for studying NE differentiation. The first study showing LNCaP cell differentiation towards an NE phenotype demonstrated that permeant analogues of cyclic AMP (cAMP) could promote neurite extension and the expression of NE markers, CgA, NSE and S100 protein (Bang et al. 1994). As in PC12 cells, NE differentiation was associated in LNCaP cells with an enhanced survival and a lower proliferation rate. NE differentiation was later shown to be induced in LNCaP cells by the depletion of steroid (Shen et al. 1997; Yuan et al. 2006) or by the addition of interleukin (IL)-1, IL-2 or IL-6 (Diaz et al. 1998; Mori et al. 1999) to the culture medium. Various neuropeptides, such as vasoactive intestinal peptide (VIP) (Collado et al. 2004, 2005; Gutierrez-Canas et al. 2005), PACAP (Farini et al. 2003), melatonin (Sainz et al. 2005) or adrenomedullin (Berenguer et al. 2008), also promote NE differentiation of LNCaP cells.

LNCaP and PC12 cells thus share many common features that make them valuable and easy-to-use cell models for studying NE differentiation. Nonetheless, they are each characterized by specific differentiation patterns. For example, NGF, which is a good inducer of NE differentiation in PC12 cells, does not promote a neuronal phenotype in prostate cancer cells (Angelsen et al. 1998).

In addition to these two cell lines, a number of other cell models have been used to study an important aspect of neuronal differentiation, i.e. neuritogenesis, notably neuroblastoma cell lines, retinoblastoma cell lines, embryonic stem cells or neurons (Tojima et al. 2000; Kushmerick et al. 2001; Hirooka et al. 2002).

4.3.2 Signalling Pathways

This topic falls beyond the purpose of this review. Therefore, only the main signalling pathways common to LNCaP and PC12 cells will be rapidly described here. Some of the pathways leading to NE differentiation are depicted in Fig. 4.1.

In both cell lines, the activation of a cAMP-dependent pathway is clearly shown to induce a neuronal phenotype. In LNCaP cells, the induction of a neuronal phenotype by stimuli raising the cytosolic cAMP concentration clearly depends on cAMP-dependent protein kinase (PKA) (Cox et al. 2000). The cAMP-dependent pathway was later shown to involve the activation of mitogen-activated protein kinase/extracellular signal-regulated kinase (MAPK/ERK) pathway (Deeble et al. 2001), which then phosphorylates downstream targets like the cAMP-responsive element-binding protein (CREB) (Farini et al. 2003). The activation of the ERK1/2 pathway by heparin-binding epidermal growth factor-like growth factor (HB-EGF) also induces NE differentiation of LNCaP cells (Kim et al. 2002). In PC12 cells, it has been shown that specific stimulation of EPAC (exchange proteins activated by cAMP) promotes neuritogenesis, whereas specific PKA activation leads to the ERK1/2 pathway and cell proliferation (Kiermayer et al. 2005). On the other hand, PACAP stimulates neurite outgrowth via ERK activation (Lazarovici et al. 1998) and the NF-κB pathway (Manecka et al. 2013). In addition, the activation by androgen removal of the PI3K–Akt–mTOR pathway may be essential to induce NE differentiation in LNCaP cells (Wu and Huang 2007), as is the case for neuritogenesis-induced NGF in PC12 cells (Jeon et al. 2010).

4.4 Alterations in Calcium Homeostasis

In addition to the signalling pathways described above, calcium ion has been proposed as a key messenger involved in NE differentiation. In LNCaP cells, it has been shown that VIP, a neuropeptide inducing the expression of NE markers, such as NSE and CgA (Collado et al. 2004; Gutierrez-Canas et al. 2005), also

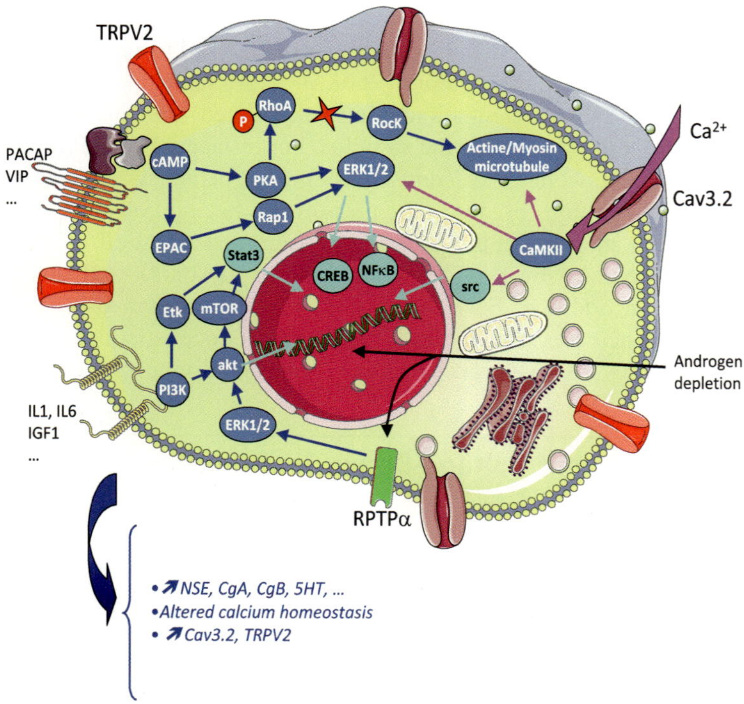

Fig. 4.1 Overview of the transduction pathways leading to NE differentiation in LNCaP cells. NE differentiation can be induced in LNCaP cells by peptides activating G protein-coupled receptors (GPCR) (PACAP, VIP, bombesin), tyrosine kinase receptors (NGF, IGF-1, IL-1, IL-6) or androgen depletion through increased expression of receptor protein-tyrosine phosphatase alpha (RPTPα) (Zhang et al. 2003). PKA–ERK–CREB (Cox et al. 2000), PI3K–Akt–mTOR (Wu and Huang 2007) and PI3K/Etk/Stat3 pathways activate NE differentiation characterized by the features described below [for a review, see Yuan et al. (2007)]. Phosphorylation of RhoA by PKA leads to its inactivation, to Rho kinase (Rock) inhibition and to neurite extension (Jones and Palmer 2012). Calcium entry through Cav3.2 channels may increase cytosolic calcium concentration, thus leading to activation of CaM kinase II and further activation of Src, leading to gene expression. CaM kinase II stimulation may also promote the cytoskeleton reorganization necessary for morphological differentiation

rapidly increases cytosolic calcium levels, c-Fos, proto-oncogenes and VEGF mRNA expression. This is associated with VEGF secretion and neurite lengthening. All these effects were shown to be antagonized by chelating intracellular calcium using BAPTA–AM (Collado et al. 2005), which therefore clearly demonstrates that NE differentiation is a calcium-dependent process. It has also been shown in the same cells that NE differentiation, induced by either steroid depletion or cytosolic cAMP elevation, is associated with a profound alteration of calcium homeostasis (Vanoverberghe et al. 2004). Indeed, NE differentiation is associated with reduced endoplasmic reticulum calcium-store content, which is attributed to decreased calreticulin (a calcium-binding/storage protein) expression and SERCA2b-calcium ATPase (Vanoverberghe et al. 2004). In addition, this has been correlated with smaller capacitative calcium entry (Vanden Abeele et al. 2003; Vanoverberghe et al. 2004, 2012; Flourakis et al. 2010) and to an increased resting cytosolic calcium concentration (Mariot et al. 2002; Monet et al. 2010).

In PC12 cells, NE differentiation is also associated with cytosolic calcium increases upon the application of the stimulus. For instance, PACAP promotes an elevation of the intracellular calcium concentration, which is an essential step towards the acquisition of the NE phenotype. Selenoprotein T (SelT), a selenocysteine, is an essential player in this calcium-dependent differentiation. Its expression is enhanced by PACAP (Grumolato et al. 2008), and this requires calcium entry. In addition, SelT is involved in calcium homeostasis since its overexpression increases the intracellular calcium concentration, whereas its knockdown reduces PACAP-induced calcium signals (Grumolato et al. 2008).

4.5 Expression of Voltage-Dependent Calcium Channels

Since NE cells are characterized by electrical excitability, voltage-dependent calcium channels have been considered as putative actors of the NE differentiation process. Voltage-dependent calcium channels are classified, based on their biophysical and pharmacological properties, as low-voltage-activated calcium channels (LVA or T-type calcium channels) and high-voltage-activated calcium channels (HVA or L-, N-, P-/Q-, R-type calcium channels). On a molecular basis, ten different α1 pore subunits have been identified that present voltage-dependent calcium channel properties (Catterall 2011): Cav1.1, Cav1.2, Cav1.3, Cav1.4 (L-type channels), Cav2.1 (P/Q channels), Cav2.2 (N type), Cav2.3 (R type) and Cav3.1, Cav3.2 and Cav3.3 (T-type channels).

Regarding the expression levels of voltage-dependent calcium channels, their involvement in neuronal or NE differentiation has been investigated in various cell models displaying the initial steps of differentiation. LVA calcium channels have thus been shown, along with HVA calcium channels, to be differentially regulated during NE or neuronal differentiation. It has been shown in the literature that in many neuronal models, T-type calcium channels are expressed in the early stages of development and may participate in the differentiation process (Holliday and Spitzer 1990; Gu and Spitzer 1993). The involvement of calcium channels, notably T-type calcium channels, in neuronal development has been reviewed thoroughly elsewhere (Lory et al. 2006).

4.5.1 Neuroblastoma Cell Lines

In the neuroblastoma X glioma hybrid NG108-15 cell line, it was shown from electrophysiological experiments and pharmacological evidence (Lukyanetz 1998) that only T-type calcium channels are expressed in the undifferentiated state. Moreover, neuronal differentiation induced by prostaglandin and isobutylmethylxanthine (IBMX) is accompanied by an overexpression of T-type calcium channels and the neo-expression of other voltage-dependent calcium channels (L-, N-, Q-type channels). These results led to the hypothesis that T-type calcium channels may be some of the channels involved in triggering neuronal differentiation (see below). SN56 cells, another hybrid cell line resulting from the fusion of neuroblastoma cells and septal neurons, express voltage-dependent calcium channels upon neuronal differentiation. In their undifferentiated state, SN56 cells mostly express T-type calcium channels (Kushmerick et al. 2001). Based on their kinetics and their pharmacological sensitivity to nickel, the authors suggested that T-type calcium currents were carried by Cav3.2 pore subunits. Upon differentiation induced by serum deprivation and cAMP, an increase in T-type calcium currents was observed within 2–4 days, together with a de novo expression of L-, N-, P-/Q- and R-type HVA calcium channels (Kushmerick et al. 2001). As in NG108-15 cells, it is therefore hypothesized that differentiation is dependent on T-type calcium channel activity in SN56 cells.

In another human neuroblastoma cell line B2(2)-M17, neuronal differentiation (evidenced by an increased expression of NSE, α7-Ach nicotinic receptor and SNAP-25) induced by retinoic acid leads to increased voltage-dependent calcium entry upon KCl depolarization (Andres et al. 2013). Based on pharmacological demonstrations, this was mainly attributed to N and P/Q channels together with a slight participation of T-type calcium channels.

4.5.2 Retinoblastoma Cells

In the human retinoblastoma Y-79 cell line, where both Cav3.1 and Cav3.2 are expressed in undifferentiated cells, there is an important reduction in mRNA levels for both channels, along with a diminution of T-type calcium current in cells undergoing neuronal differentiation (Hirooka et al. 2002). Downregulation of Cav3.1 T-type channel activity and expression was further demonstrated to be due to a reduced activity of the promoters controlling its transcriptional expression (Bertolesi et al. 2003). A more recent study analysing modifications in alternative splicing showed that there is a considerable change in Cav3.1 splice variants occurring during neuronal differentiation (Bertolesi et al. 2006). Nevertheless, knocking down the expression of Cav3.1 failed to alter differentiation kinetics or neurite formation. These results led the authors to suggest that Cav3.1 and Cav3.2 channels are not involved in the differentiation process itself but that the alterations in splice variant-specific expression may be relevant for establishing a mature differentiated state (Hirooka et al. 2002).

4.5.3 Embryonic Stem Cells

In embryonic stem (ES) cells, neuronal differentiation promotes electrical excitability and is associated with maturation of voltage-dependent calcium channels (Arnhold et al. 2000). This differentiation can be induced by various neuropeptides, including VIP or PACAP. Both peptides stimulate neurite outgrowth and NSE overexpression after 4–8 days of treatment (Cazillis et al. 2004). This is associated with an enhancement of T-type calcium currents (Chafai et al. 2011) correlated with a stronger expression of Cav3.3 protein as observed in Western blot experiments.

4.5.4 NE Cell Lines

It has been shown in PC12 cells that NE differentiation is accompanied by increased expression of both LVA (Garber et al. 1989) and HVA calcium channels (Shitaka et al. 1996; Sherwood et al. 1997). Chronic application of various agents that induce NE differentiation, such as NGF or dexamethasone, leads to an increase in both the proportion of cells expressing LVA calcium currents and the average LVA calcium current density (Garber et al. 1989). NE differentiation induced by PACAP or NGF is associated with increased density of voltage-dependent calcium currents, both of the LVA and the HVA types (Grumolato et al. 2003).

In the NE pituitary lactosomatotrope GH3 cell line, further differentiation towards a lactotrope phenotype may be induced by treatment with NGF for several days. While a short treatment with NGF only upregulates HVA calcium channels, a sustained NGF action (1–5 days), occurring through a p75 receptor, leads to upregulated prolactin mRNA and protein levels and to a delayed increase in T-type calcium currents (Lopez-Dominguez et al. 2006). These T-type calcium currents are possibly carried by Cav3.1 and/or Cav3.3 channels since they are the only LVA subunits expressed in GH3 cells (Mudado et al. 2004).

In the LNCaP prostate cancer cell line, previously shown to differentiate following cAMP stimuli (Bang et al. 1994), it was demonstrated that 3–4 days after the onset of NE differentiation, there was an increase in the magnitude of voltage-dependent calcium currents. This calcium current displayed all the characteristics of T-type calcium channels (Mariot et al. 2002). Indeed, it was characterized by a fast and voltage-dependent inactivation and a slow deactivation as well as being inhibited by low concentrations of $NiCl_2$, mibefradil, flunarizine and kurtoxin. The increase in T-type calcium current took longer to occur than the lengthening of neuritic extension which developed within a few hours of the early steps of differentiation. It was demonstrated by quantitative RT-PCR that the enhancement of T-type calcium currents depends on an overexpression of Cav3.2 channels. A particular feature of LNCaP cells is that Cav3.2 calcium channels are the only voltage-dependent calcium channels expressed in both undifferentiated and differentiated conditions. Cav3.2 overexpression was associated with an increase in resting cytosolic calcium concentration, antagonized by T-type calcium channel inhibitors (Mariot et al. 2002). It was shown that T-type calcium channels, due to their combined inactivation and activation properties, participate in basal calcium entry at a resting membrane potential of 40 mV in LNCaP prostate cells (Gackiere et al. 2006).

4.6 Role of T-Type Calcium Channels in NE Differentiation

There has been some debate concerning the positive or adverse effects of intracellular calcium and voltage-dependent calcium channels on neuritic extension. While some published studies show no role for voltage-dependent calcium channels in neurite extension induced by NGF (Lichvarova et al. 2012), most publications report a stimulation of neurite outgrowth by increased cytosolic calcium levels due to calcium entry through several channels.

Indeed, in many cell models, calcium is necessary for neurite outgrowth. For instance, it has recently been shown in PC12 cells that PACAP is able to induce neuritogenesis through a NF-κB pathway. The authors demonstrated that PACAP-induced NE differentiation occurs through the activation of a calcium-dependent ERK1/2 MAP kinase pathway and the recruitment of cRel and p52 subunits of NF-κB, leading to neuritogenesis and cell survival (Manecka et al. 2013). In that study, neuritogenesis and cRel recruitment are both inhibited by non-specific inhibitors of plasma membrane calcium channels and intracellular calcium channels responsible for endoplasmic reticulum calcium release. In PC12 cells, calcium homeostasis perturbation, for example, induced by knocking down the expression of plasma membrane Ca^{2+}-ATPase (PMCA 2 and 3), leads to alterations in the process of morphological differentiation characterized by neuritic extension (Boczek et al. 2012).

T-type calcium channels have thus been investigated as putative key players of the differentiation process since, as described above, they are frequently expressed in either the early or late stages of NE/neuronal differentiation.

4.6.1 Role in the Expression of Molecular NE Markers

To our knowledge, it has never been demonstrated that T-type calcium channel activity or expression promotes the expression of molecular NE markers such as chromogranins or NSE. We show here (Fig. 4.2a and b—personal observations) that overexpressing Cav3.2 T-type calcium channels in LNCaP cells induces increased expression of NSE and CgA. In addition, NE differentiation induced by activating the PKA pathway (using a combined treatment with permeant cyclic AMP analogues and IBMX) leads to elevated NSE expression which is antagonized by siRNA against Cav3.2 (Fig. 4.2b). This shows that Cav3.2 channel expression is not only increased during NE differentiation process but also that it plays a central role in some aspects of the differentiation process.

4.6.2 Role in Neuritic Extension

The role of T-type calcium channels in neurite formation has been investigated in embryonic neural progenitor cells (Louhivuori et al. 2013). From the inhibition of calcium signals induced mainly by T-type channel inhibitors (mibefradil,

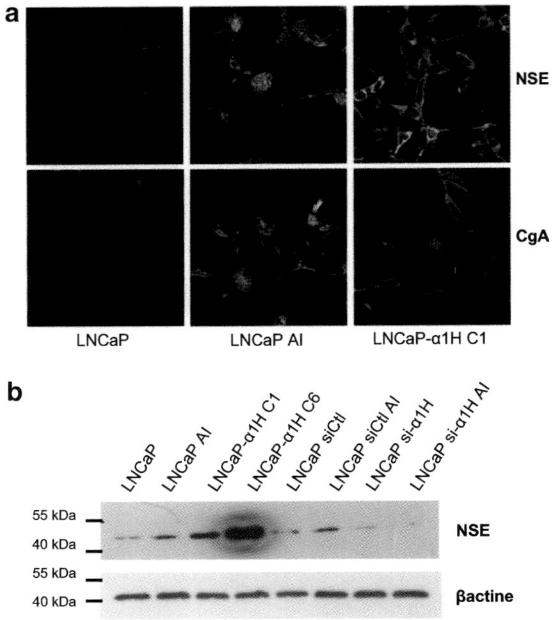

Fig. 4.2 Cav3.2 channels enhance the expression of NE markers in LNCaP. (**a**) Immunofluorescence studies using NSE and CgA-specific antibodies (dilution 1/50e, DAKO, France) on LNCaP cells. The cells were treated with a permeant analogue of cyclic AMP (dibutyryl cyclic AMP, 1 mM) and isobutylmethylxanthine (IBMX, 100 μM) for 4 days to induce NE differentiation (LNCaP-AI). NSE and CgA staining (in *green*) were increased by the activation of the PKA pathway or the overexpression of Cav3.2 channels (LNCaP-α1H C1). Nuclei are stained in *blue* using DAPI. (**b**) Western blot experiments performed on LNCaP-Ctl, LNCaP-AI, LNCaP-α1H (2 clones: C1 and C6) and the same cells with siRNA treatment. Cells were transfected overnight with control siRNA (siCtl 50 nM) or siRNA targeting Cav3.2 (siα1H 50 nM—see Gackiere et al. (2008) for siRNA sequences). The cells were harvested 4 days after transfection, and a Western blot experiment was performed using the same antibody as the immunofluorescence studies (except for β-actin, dilution 1/2000e, Sigma)

NNC-550396), it was concluded in this study that T-type calcium channels are expressed during the early stages of neuronal differentiation (1 day). However, during the differentiation process, there is an increase in the density of HVA channels compared to LVA channels, such that at differentiation day 20, calcium signals are totally inhibited by nifedipine. The authors showed that inhibitors of LVA channels significantly decrease the number of active migrating neuron-like cells and neurite extensions, thereby demonstrating their involvement in neuritogenesis.

Early reports on neuroblastoma cells demonstrated that the expression of T-type currents always precedes neurite extension, thus suggesting that calcium entry through T-type calcium channels may participate in morphological neuronal differentiation (Silver and Bolsover 1991).

In the NG108-15 cell line, it was shown that cAMP-induced neuronal differentiation, characterized by neurite formation and an overexpression of L-type calcium channels, is dependent on T-type calcium channels activity. Indeed, the inhibition of T-type calcium channels using $NiCl_2$ or the downregulation of the Cav3.2 pore subunit led to a decrease in the number of cells with neurite, but did not change the average neurite length (Chemin et al. 2002). In addition, Cav3.2 T-type calcium channels were necessary players in HVA calcium channel expression during neuronal differentiation. This led the authors to propose that Cav3.2 T-type calcium channels are an early actor in the neuronal differentiation process.

In prostate cancer LNCaP cells, neurite formation induced by cAMP is faster (occurring in few hours) than Cav3.2 overexpression (occurring in few days), which suggests that Cav3.2 overexpression may only be a consequence of the differentiation process. However, there is significant expression of Cav3.2 channels and T-type calcium current, in the undifferentiated state which could be responsible for the morphological differentiation process. We have indeed shown that neurite elongation during NE differentiation is dependent on Cav3.2 calcium channels since their inhibition by low $NiCl_2$ concentrations reduces the average neurite length without hampering neurite formation. Therefore, if neurite formation is not dependent on Cav3.2 channels, their lengthening is modulated by T-type calcium channel activity (Mariot et al. 2002).

4.6.3 Role in Hormone or Neurohormone Secretion

The central role of cytosolic calcium in the regulated pathway of secretion (and therefore in hormones, neurohormones and neurotransmitter release) has been demonstrated in many endocrine, NE or neuronal cells. The role of HVA channels in regulated exocytosis has been known for a long time, but attention has been paid more recently to LVA channels, and their role in calcium-dependent secretion or neurotransmission has now been documented (Carbone et al. 2006).

Anterior NE pituitary cells express both LVA and HVA calcium channels (Lewis et al. 1988), and although HVA channels regulate hormone secretion, pharmacological experiments on perifused rat pituitaries have demonstrated that T-type calcium channels also participate in the secretion of hypophyseal hormones like thyrotropin, in response to hypothalamic neuropeptides (Roussel et al. 1992). In addition, in NE pituitary cells, the expression of Cav3.1 is raised, whereas the expression of Cav3.2 and Cav3.3 is reduced in mice treated with 17ß-estradiol. This regulation of T-type calcium channel mRNA expression could be an important mechanism involved in the 17ß-estradiol regulation of pituitary secretion (Bosch et al. 2009). A convincing link has been demonstrated for L-type calcium channels and exocytosis in many NE cell models. However, it has been shown in various NE cells, including rat pituitary melanotropes, that T-type calcium channels are as efficient as L-type channels in promoting exocytosis. In these cells, calcium ions "couple with equal strength to exocytosis regardless of the channel type involved" (Mansvelder and Kits 2000).

In prostate NE LNCaP cells, we have shown that membrane depolarization, due to a stimulation of T-type calcium current, induces calcium-dependent exocytosis, as measured with FM1-43 dye. In addition, we showed that prostatic acid phosphatase (PAP) secretion is calcium-dependent in prostate cells and that this secretion is potentiated by NE differentiation. Furthermore, PAP secretion is reduced by low concentrations of $NiCl_2$, flunarizine or kurtoxin and by siRNAs targeting Cav3.2 channels. Cav3.2 expression and activity are therefore responsible for PAP secretion in NE prostate cancer cells (Gackiere et al. 2008).

In adult chromaffin cells, the expression of LVA channels is weak, and HVA channels control to a large extent voltage-dependent calcium influx and thereby catecholamine secretion (Artalejo et al. 1994). However, the expression of LVA channels is more extensive in embryonic chromaffin cells (Bournaud et al. 2001) than in adult cells. It therefore seems that during development, chromaffin cells lose the capacity to express LVA channels, which are progressively replaced by HVA channels. In adult cells, the Cav3.2 isoform is weakly expressed but is upregulated either by a treatment that increases cytosolic cAMP concentration (Novara et al. 2004) or by chronic hypoxia (Carabelli et al. 2007b). This cAMP effect was demonstrated to be mediated by a PKA-independent signalling pathway through the cAMP-receptor protein Epac (cAMP–guanine nucleotide exchange factor (cAMP–GEF)) (Novara et al. 2004). In addition, newly formed Cav3.2 channels are as efficient as HVA channels in promoting low-threshold exocytosis and the release of catecholamine (Giancippoli et al. 2006; Carabelli et al. 2007a, b). In mouse chromaffin cells in situ, short-term exposure to PACAP stimulates T-type calcium channel activity through a PKC-dependent recruitment of Cav3.2 channels to the plasma membrane (Hill et al. 2011). Cav3.2 channels may therefore be responsible for catecholamine secretion in acute stress situations.

It has thus been widely shown that LVA calcium channels control secretion and neurotransmission. This has been discussed in a recent review (Weiss and Zamponi 2013). However, the mechanisms involved in that function have as yet remained unclear. To promote a fast release of neuropeptides or neurotransmitters, HVA channels, for their part, are localized in close vicinity of the vesicle-release machinery. It has been demonstrated that this co-localization is achieved using a specific interaction between HVA channels (Cav2.1 and Cav2.2) and SNARE proteins (syntaxin-1, SNAP-25 and synaptotagmin-1), through an interaction site called a synprint. It has recently been shown that although T-type calcium channels lack this synprint interaction site, they closely associate with syntaxin-1A in central neurons (Weiss et al. 2012). This association between syntaxin-1A and Cav3.2 was also demonstrated in the same study to be essential for low-threshold exocytosis in chromaffin cells.

4.7 How Do T-Type Calcium Channels Mediate NE Differentiation? Putative Signalling Mechanisms

The downstream events after calcium entry through T-type calcium channels in the context of NE differentiation have been rarely studied. Since T-type calcium channels have been shown to be involved in secretion, it has been investigated whether their role in the induction of an NE phenotype could be related to the secretion of growth factors. As shown by one study (Chemin et al. 2004), neuritogenesis could be induced in NG108-15 cells using the conditioned media from Cav3.2 expressing NG108-15 cells. When the expression of Cav3.2 channels was knocked down with siRNAs, conditioned media from these siRNA-treated cells did not induce neurite outgrowth in NG108-15 cells, thus showing the involvement of T-type channels in a paracrine loop promoting neurite formation. However, the nature of the factors secreted by NG108-15 cells in response to T-type channels activity has not been elucidated.

In addition, T-type calcium channels may promote NE or neuronal differentiation through a stimulation of gene expression. It has been shown in NG108-15 cells that T-type calcium channels are responsible for the induction by hydrogen sulphide (NaHS) of a neuronal phenotype, characterized by the overexpression of L-type calcium channels and neuritogenesis (Nagasawa et al. 2009). This was essentially demonstrated using T-type calcium channel inhibitors, such as mibefradil or zinc chloride in a range of concentrations more specifically inhibiting Cav3.2 channels. In the same cell line, it was later demonstrated that NaHS promotes neuritic extension through a phosphorylation of Src kinase (Tarui et al. 2010). Indeed, PP2, a specific inhibitor of Src, antagonized the development of neurites induced by NaHS, while only reducing the ones activated by permeant cAMP. In addition, Src phosphorylation by NaHS was significantly reduced by inhibitors of T-type calcium channels, mibefradil or ascorbate, showing that T-type calcium channels are upstream of Src in the pathway leading to a neuronal phenotype. It was shown in this study that Src phosphorylation is independent of CaM Kinase II. Another study demonstrated that depolarization-induced calcium entry promotes NE differentiation of PC12 cells. This NE differentiation, characterized by neurite formation, GAP-43 and synapsin I overexpression, requires the activation of phospholipase D_2–ErK–CREB pathway and Src phosphorylation by CaM Kinase II (Banno et al. 2008). Such a pathway is similar to the one demonstrated in hippocampal neurons, where NCAM (neural cell adhesion molecules)-induced neurite formation is dependent on the T-type calcium channel (Kiryushko et al. 2006). In these cells, homophilic NCAM interaction, which promotes neurite formation, leads to an increase in cytosolic calcium via T-type calcium channels, together with L-type and non-selective cation channels, and to the activation of an Src-dependent pathway, both events being necessary for the induction of differentiation.

4.8 Other Calcium Channels

Though T-type calcium channels are indeed involved in neuronal and NE differentiation as previously exemplified, other calcium channels are part of that process. In PC12 cells, L-type calcium channel stimulation, through 5-HT3 receptor activation by serotonin, has been shown to potentiate NGF-induced neurite outgrowth (Homma et al. 2006), due to increased intracellular calcium concentration. Similarly, lysophosphatidylcholine promotes neurite outgrowth in an L-type channel-dependent manner (Nakashima et al. 2003). In the same cells, NGF-induced neurite outgrowth is correlated with changes in the expression pattern of TRPC (transient receptor potential canonical) channels, TRPC1 being overexpressed and TRPC5 downregulated (Kumar et al. 2012). Furthermore, TRPC1 stimulates neurite outgrowth, whereas TRPC5 inhibits it (Heo et al. 2012; Kumar et al. 2012). A similar inhibitory action of TRPC5 was observed in hippocampal neurons, where the overexpression of dominant negative TRPC5 channels increases neurite length (Greka et al. 2003). In LNCaP prostate cancer cells, NE differentiation is associated not only with an upregulation of Cav3.2 (Mariot et al. 2002) but also with TRPV2 (TRP vanilloid) overexpression (Monet et al. 2010) and with a downregulation of Orai1 (Flourakis et al. 2010; Vanoverberghe et al. 2012). Overall, calcium homeostasis is therefore deeply altered by treatments that induce NE differentiation, leading to an increase in basal cytosolic concentration, reduced calcium-store content and reduced capacitative calcium entry.

4.9 NE Differentiation in Cancer Progression: Perspectives on the General Involvement of T-Type Calcium Channels

NE tumours may arise from the proliferation of NE cells. These tumours develop in almost all tissues, including those where NE cells may not be present (DeLellis 2001). It must, however, be noted that the digestive system is the most frequent site of development of NE tumours (66 %) followed by the respiratory tract (31 %) (Gustafsson et al. 2008). NET, although slow growing, are often diagnosed after the tumour has metastasized and thus may entail a poor prognosis. Some tissues or organs may also be the site for the development of carcinoma that displays NE cell foci, which has led to the notion of cancers with NE differentiation. This is the case for prostate cancer, which is frequently characterized by NE differentiation (Abrahamsson 1999; Hansson and Abrahamsson 2003). In prostate cancer, NE differentiation is usually correlated with a poor prognosis (Sagnak et al. 2011), probably due to the fact that NE cells are androgeno-insensitive (Huang et al. 2006) and may thus participate in the androgen-independent growth of prostate cancer. In addition, these cells are resistant to apoptosis (Vanoverberghe et al. 2004) and secrete numerous factors that may increase cell proliferation in their vicinity. In some forms of prostate carcinoma, NE cells are the main component of the tumours. These small-cell prostate carcinomas (SCPCa) are very aggressive forms of prostate cancer, where NE cells exhibit tumorigenic and highly proliferative activities

(Oesterling et al. 1992; Yao et al. 2006). As shown recently, SCPCa cells may arise from NE carcinoma cells from a p53-inactivating mutation, a mutation conferring proliferative properties (Chen et al. 2012).

As reported in a number of cancers (for a review, see Panner and Wurster 2006), T-type calcium channels may participate in tumour growth. Their role in cell proliferation may be due to the fact that enhanced expression of T-type calcium channels (Cav3.1, Cav3.2 or Cav3.3) leads to increased secretion of tumorigenic factors by NE cells. As we have previously shown, PAP synthesis and release are enhanced by Cav3.2 channels in LNCaP cells (Gackiere et al. 2008). In addition, the secreted form of PAP may play an important role in the development of prostate tumour metastasis. Indeed, PAP expression is increased in prostate bone metastasis, and PAP may participate in the osteoblastic phase of the metastasis development (Kirschenbaum et al. 2011). Moreover, we have shown that the expression of Cav3.2 channels is correlated with the expression of serotonin (Gackiere et al. 2008), a neurotransmitter stimulating the proliferation and migration of prostate cancer cell lines (Dizeyi et al. 2011). Furthermore, we have shown that Cav3.2 expression correlates with CgA expression (Gackiere et al. 2008). The circulating level of CgA, an NE marker, is frequently increased in the advanced stages of the diseases. It has been demonstrated in prostate cancer cell lines that CgA accelerates cell proliferation and reduces cell apoptosis, thereby suggesting a role for CgA in prostate cancer development (Gong et al. 2007). Altogether, this suggests that Cav3.2 channels in NE prostate cancer cells may promote calcium entry, which in turn induces the secretion of mitogenic or tumorigenic factors. In addition, we have observed that Cav3.2 channels in prostate cancer LNCaP cells favour cell growth, even in NE cells, since their overexpression, which increases NSE expression, also accelerates cell proliferation. In contrast, downregulation of Cav3.2, which reduces NSE expression, or its blockade with various inhibitors, slows cell proliferation (Gackiere et al. 2013). This is in agreement with many articles showing the participation of T-type calcium channels in various aspects of cell growth, including cell proliferation or cell apoptosis [for a review, see Panner and Wurster (2006) or Lory et al. (2006)]. Since NE cells of the prostate may proliferate rapidly in some forms of prostate carcinomas (SCPCa), it will be of great interest to assess whether Cav3.2 channels, involved in both proliferation and NE differentiation, in combination with a p53 inactivation, are indeed involved in the development of SCPCa.

Concluding Remarks

LVA calcium channels have frequently been shown to be overexpressed during NE differentiation as well as being involved in the development of an NE phenotype characterized by morphological (neurite extension) and molecular (NSE or CgA markers) features. Moreover, T-type calcium channels, along with other calcium channels, participate in the secretion of paracrine or endocrine factors, which in turn may promote the differentiation or the proliferation of target cells. In the context of NE tumours, most of the data available rely on cell line studies. It is of particular importance in the near future to carry out major studies on tissue arrays, in order to correlate the stage of the disease with the expression of the different T-type calcium channel isoforms.

References

Abrahamsson PA (1999) Neuroendocrine differentiation in prostatic carcinoma. Prostate 39:135–148

Andres D, Keyser BM, Petrali J, Benton B, Hubbard KS, McNutt PM, Ray R (2013) Morphological and functional differentiation in BE(2)-M17 human neuroblastoma cells by treatment with Trans-retinoic acid. BMC Neurosci 14:49

Andrew A (1974) Further evidence that enterochromaffin cells are not derived from the neural crest. J Embryol Exp Morphol 31:589–598

Angelsen A, Sandvik AK, Syversen U, Stridsberg M, Waldum HL (1998) NGF-beta, NE-cells and prostatic cancer cell lines. A study of neuroendocrine expression in the human prostatic cancer cell lines DU-145, PC-3, LNCaP, and TSU-pr1 following stimulation of the nerve growth factor-beta. Scand J Urol Nephrol 32:7–13

Arnhold S, Andressen C, Angelov DN, Vajna R, Volsen SG, Hescheler J, Addicks K (2000) Embryonic stem-cell derived neurones express a maturation dependent pattern of voltage-gated calcium channels and calcium-binding proteins. Int J Dev Neurosci 18:201–212

Artalejo CR, Adams ME, Fox AP (1994) Three types of Ca^{2+} channel trigger secretion with different efficacies in chromaffin cells. Nature 367:72–76

Bang YJ, Pirnia F, Fang WG, Kang WK, Sartor O, Whitesell L, Ha MJ, Tsokos M, Sheahan MD, Nguyen P et al (1994) Terminal neuroendocrine differentiation of human prostate carcinoma cells in response to increased intracellular cyclic AMP. Proc Natl Acad Sci USA 91:5330–5334

Banno Y, Nemoto S, Murakami M, Kimura M, Ueno Y, Ohguchi K, Hara A, Okano Y, Kitade Y, Onozuka M, Murate T, Nozawa Y (2008) Depolarization-induced differentiation of PC12 cells is mediated by phospholipase D2 through the transcription factor CREB pathway. J Neurochem 104:1372–1386

Berenguer C, Boudouresque F, Dussert C, Daniel L, Muracciole X, Grino M, Rossi D, Mabrouk K, Figarella-Branger D, Martin PM, Ouafik L (2008) Adrenomedullin, an autocrine/paracrine factor induced by androgen withdrawal, stimulates 'neuroendocrine phenotype' in LNCaP prostate tumor cells. Oncogene 27:506–518

Bergeron F, Leduc R, Day R (2000) Subtilase-like pro-protein convertases: from molecular specificity to therapeutic applications. J Mol Endocrinol 24:1–22

Bertolesi GE, Jollimore CA, Shi C, Elbaum L, Denovan-Wright EM, Barnes S, Kelly ME (2003) Regulation of alpha1G T-type calcium channel gene (CACNA1G) expression during neuronal differentiation. Eur J Neurosci 17:1802–1810

Bertolesi GE, Walia Da Silva R, Jollimore CA, Shi C, Barnes S, Kelly ME (2006) Ca(v)3.1 splice variant expression during neuronal differentiation of Y-79 retinoblastoma cells. Neuroscience 141:259–268

Boczek T, Lisek M, Kowalski A, Pikula S, Niewiarowska J, Wiktorska M, Zylinska L (2012) Downregulation of PMCA2 or PMCA3 reorganizes $Ca^{(2+)}$ handling systems in differentiating PC12 cells. Cell Calcium 52:433–444

Bosch MA, Hou J, Fang Y, Kelly MJ, Ronnekleiv OK (2009) 17Beta-estradiol regulation of the mRNA expression of T-type calcium channel subunits: role of estrogen receptor alpha and estrogen receptor beta. J Comp Neurol 512:347–358

Bournaud R, Hidalgo J, Yu H, Jaimovich E, Shimahara T (2001) Low threshold T-type calcium current in rat embryonic chromaffin cells. J Physiol 537:35–44

Carabelli V, Marcantoni A, Comunanza V, Carbone E (2007a) Fast exocytosis mediated by T- and L-type channels in chromaffin cells: distinct voltage-dependence but similar Ca^{2+}-dependence. Eur Biophys J 36:753–762

Carabelli V, Marcantoni A, Comunanza V, de Luca A, Diaz J, Borges R, Carbone E (2007b) Chronic hypoxia up-regulates alpha1H T-type channels and low-threshold catecholamine secretion in rat chromaffin cells. J Physiol 584:149–165

Carbone E, Giancippoli A, Marcantoni A, Guido D, Carabelli V (2006) A new role for T-type channels in fast "low-threshold" exocytosis. Cell Calcium 40:147–154

Catterall WA (2011) Voltage-gated calcium channels. Cold Spring Harb Perspect Biol 3:a003947

Cazillis M, Gonzalez BJ, Billardon C, Lombet A, Fraichard A, Samarut J, Gressens P, Vaudry H, Rostene W (2004) VIP and PACAP induce selective neuronal differentiation of mouse embryonic stem cells. Eur J Neurosci 19:798–808

Chafai M, Basille M, Galas L, Rostene W, Gressens P, Vaudry H, Gonzalez BJ, Louiset E (2011) Pituitary adenylate cyclase-activating polypeptide and vasoactive intestinal polypeptide promote the genesis of calcium currents in differentiating mouse embryonic stem cells. Neuroscience 199:103–115

Chemin J, Nargeot J, Lory P (2002) Neuronal T-type alpha 1H calcium channels induce neuritogenesis and expression of high-voltage-activated calcium channels in the NG108-15 cell line. J Neurosci 22:6856–6862

Chemin J, Nargeot J, Lory P (2004) Ca(v)3.2 calcium channels control an autocrine mechanism that promotes neuroblastoma cell differentiation. Neuroreport 15:671–675

Chen H, Sun Y, Wu C, Magyar CE, Li X, Cheng L, Yao JL, Shen S, Osunkoya AO, Liang C, Huang J (2012) Pathogenesis of prostatic small cell carcinoma involves the inactivation of the P53 pathway. Endocr Relat Cancer 19:321–331

Collado B, Gutierrez-Canas I, Rodriguez-Henche N, Prieto JC, Carmena MJ (2004) Vasoactive intestinal peptide increases vascular endothelial growth factor expression and neuroendocrine differentiation in human prostate cancer LNCaP cells. Regul Pept 119:69–75

Collado B, Sanchez MG, Diaz-Laviada I, Prieto JC, Carmena MJ (2005) Vasoactive intestinal peptide (VIP) induces c-fos expression in LNCaP prostate cancer cells through a mechanism that involves Ca^{2+} signalling. Implications in angiogenesis and neuroendocrine differentiation. Biochim Biophys Acta 1744:224–233

Cox ME, Deeble PD, Bissonette EA, Parsons SJ (2000) Activated 3',5'-cyclic AMP-dependent protein kinase is sufficient to induce neuroendocrine-like differentiation of the LNCaP prostate tumor cell line. J Biol Chem 275:13812–13818

Day R, Salzet M (2002) The neuroendocrine phenotype, cellular plasticity, and the search for genetic switches: redefining the diffuse neuroendocrine system. Neuro Endocrinol Lett 23:447–451

Deeble PD, Murphy DJ, Parsons SJ, Cox ME (2001) Interleukin-6- and cyclic AMP-mediated signaling potentiates neuroendocrine differentiation of LNCaP prostate tumor cells. Mol Cell Biol 21:8471–8482

DeLellis RA (2001) The neuroendocrine system and its tumors: an overview. Am J Clin Pathol 115 (Suppl):S5–S16

Diaz M, Abdul M, Hoosein N (1998) Modulation of neuroendocrine differentiation in prostate cancer by interleukin-1 and -2. Prostate Suppl 8:32–36

Dizeyi N, Hedlund P, Bjartell A, Tinzl M, Austild-Tasken K, Abrahamsson PA (2011) Serotonin activates MAP kinase and PI3K/Akt signaling pathways in prostate cancer cell lines. Urol Oncol 29:436–445

Elhamdani A, Brown ME, Artalejo CR, Palfrey HC (2000) Enhancement of the dense-core vesicle secretory cycle by glucocorticoid differentiation of PC12 cells: characteristics of rapid exocytosis and endocytosis. J Neurosci 20:2495–2503

Farini D, Puglianiello A, Mammi C, Siracusa G, Moretti C (2003) Dual effect of pituitary adenylate cyclase activating polypeptide on prostate tumor LNCaP cells: short- and long-term exposure affect proliferation and neuroendocrine differentiation. Endocrinology 144:1631–1643

Feyrter F (1938) Uber diffuse endokrine epitheliale Organe. JA Barth, Leipzig, Germany

Flourakis M, Lehen'kyi V, Beck B, Raphael M, Vandenberghe M, Abeele FV, Roudbaraki M, Lepage G, Mauroy B, Romanin C, Shuba Y, Skryma R, Prevarskaya N (2010) Orai1 contributes to the establishment of an apoptosis-resistant phenotype in prostate cancer cells. Cell Death Dis 1:e75

Froelich F (1949) Die "Helle Zelle" der Bronchialschleimhaut und ihre Beziehungen zum Problem der Chemoreceptoren. Frankfurt Z Pathol 60:517–559

Fujita T, Kobayashi S, Yui R (1980) Paraneuron concept and its current implications. Adv Biochem Psychopharmacol 25:321–325

Gackiere F, Bidaux G, Delcourt P, Van Coppenolle F, Katsogiannou M, Dewailly E, Bavencoffe A, Van Chuoi-Mariot MT, Mauroy B, Prevarskaya N, Mariot P (2008) $Ca_V3.2$T-type calcium channels are involved in calcium-dependent secretion of neuroendocrine prostate cancer cells. J Biol Chem 283:10162–10173

Gackiere F, Bidaux G, Lory P, Prevarskaya N, Mariot P (2006) A role for voltage gated T-type calcium channels in mediating "capacitative" calcium entry? Cell Calcium 39:357–366

Gackiere F, Warnier M, Katsogiannou M, Derouiche S, Delcourt P, Dewailly E, Slomianny C, Humez S, Prevarskaya N, Roudbaraki M, Mariot P (2013) Functional coupling between large-conductance potassium channels and Cav3.2 voltage-dependent calcium channels participates in prostate cancer cell growth. Biol Open 2:941–991

Garber SS, Hoshi T, Aldrich RW (1989) Regulation of ionic currents in pheochromocytoma cells by nerve growth factor and dexamethasone. J Neurosci 9:3976–3987

Giancippoli A, Novara M, de Luca A, Baldelli P, Marcantoni A, Carbone E, Carabelli V (2006) Low-threshold exocytosis induced by cAMP-recruited $Ca_V3.2$ (alpha1H) channels in rat chromaffin cells. Biophys J 90:1830–1841

Greene LA, Tischler AS (1976) Establishment of a noradrenergic clonal line of rat adrenal pheochromocytoma cells which respond to nerve growth factor. Proc Natl Acad Sci USA 73:2424–2428

Greka A, Navarro B, Oancea E, Duggan A, Clapham DE (2003) TRPC5 is a regulator of hippocampal neurite length and growth cone morphology. Nat Neurosci 6:837–845

Grumolato L, Elkahloun AG, Ghzili H, Alexandre D, Coulouarn C, Yon L, Salier JP, Eiden LE, Fournier A, Vaudry H, Anouar Y (2003) Microarray and suppression subtractive hybridization analyses of gene expression in pheochromocytoma cells reveal pleiotropic effects of pituitary adenylate cyclase-activating polypeptide on cell proliferation, survival, and adhesion. Endocrinology 144:2368–2379

Grumolato L, Ghzili H, Montero-Hadjadje M, Gasman S, Lesage J, Tanguy Y, Galas L, Ait-Ali D, Leprince J, Guerineau NC, Elkahloun AG, Fournier A, Vieau D, Vaudry H, Anouar Y (2008) Selenoprotein T is a PACAP-regulated gene involved in intracellular Ca2+ mobilization and neuroendocrine secretion. FASEB J 22:1756–1768

Gong J, Lee J, Akio H, Schlegel PN, Shen R (2007) Attenuation of apoptosis by chromogranin A-induced Akt and survivin pathways in prostate cancer cells. Endocrinology 148:4489–4499

Gu X, Spitzer NC (1993) Low-threshold Ca^{2+} current and its role in spontaneous elevations of intracellular Ca^{2+} in developing Xenopus neurons. J Neurosci 13:4936–4948

Gustafsson BI, Kidd M, Modlin IM (2008) Neuroendocrine tumors of the diffuse neuroendocrine system. Curr Opin Oncol 20:1–12

Gutierrez-Canas I, Juarranz MG, Collado B, Rodriguez-Henche N, Chiloeches A, Prieto JC, Carmena MJ (2005) Vasoactive intestinal peptide induces neuroendocrine differentiation in the LNCaP prostate cancer cell line through PKA, ERK, and PI3K. Prostate 63:44–55

Hansson J, Abrahamsson PA (2003) Neuroendocrine differentiation in prostatic carcinoma. Scand J Urol Nephrol Suppl 28–36

Heo DK, Chung WY, Park HW, Yuan JP, Lee MG, Kim JY (2012) Opposite regulatory effects of TRPC1 and TRPC5 on neurite outgrowth in PC12 cells. Cell Signal 24:899–906

Hill J, Chan SA, Kuri B, Smith C (2011) Pituitary adenylate cyclase-activating peptide (PACAP) recruits low voltage-activated T-type calcium influx under acute sympathetic stimulation in mouse adrenal chromaffin cells. J Biol Chem 286:42459–42469

Hirooka K, Bertolesi GE, Kelly ME, Denovan-Wright EM, Sun X, Hamid J, Zamponi GW, Juhasz AE, Haynes LW, Barnes S (2002) T-Type calcium channel alpha1G and alpha1H subunits in human retinoblastoma cells and their loss after differentiation. J Neurophysiol 88:196–205

Holliday J, Spitzer NC (1990) Spontaneous calcium influx and its roles in differentiation of spinal neurons in culture. Dev Biol 141:13–23

Homma K, Kitamura Y, Ogawa H, Oka K (2006) Serotonin induces the increase in intracellular Ca2+ that enhances neurite outgrowth in PC12 cells via activation of 5-HT3 receptors and voltage-gated calcium channels. J Neurosci Res 84:316–325

Huang J, Yao JL, di Sant'Agnese PA, Yang Q, Bourne PA, Na Y (2006) Immunohistochemical characterization of neuroendocrine cells in prostate cancer. Prostate 66:1399–1406

Jeon S, Park JK, Bae CD, Park J (2010) NGF-induced moesin phosphorylation is mediated by the PI3K, Rac1 and Akt and required for neurite formation in PC12 cells. Neurochem Int 56:810–818

Jones SE, Palmer TM (2012) Protein kinase A-mediated phosphorylation of RhoA on serine 188 triggers the rapid induction of a neuroendocrine-like phenotype in prostate cancer epithelial cells. Cell Signal 24:1504–1514

Kiermayer S, Biondi RM, Imig J, Plotz G, Haupenthal J, Zeuzem S, Piiper A (2005) Epac activation converts cAMP from a proliferative into a differentiation signal in PC12 cells. Mol Biol Cell 16:5639–5648

Kim J, Adam RM, Freeman MR (2002) Activation of the Erk mitogen-activated protein kinase pathway stimulates neuroendocrine differentiation in LNCaP cells independently of cell cycle withdrawal and STAT3 phosphorylation. Cancer Res 62:1549–1554

Kirschenbaum A, Liu XH, Yao S, Leiter A, Levine AC (2011) Prostatic acid phosphatase is expressed in human prostate cancer bone metastases and promotes osteoblast differentiation. Ann N Y Acad Sci 1237:64–70

Kiryushko D, Korshunova I, Berezin V, Bock E (2006) Neural cell adhesion molecule induces intracellular signaling via multiple mechanisms of Ca^{2+} homeostasis. Mol Biol Cell 17:2278–2286

Kumar S, Chakraborty S, Barbosa C, Brustovetsky T, Brustovetsky N, Obukhov AG (2012) Mechanisms controlling neurite outgrowth in a pheochromocytoma cell line: the role of TRPC channels. J Cell Physiol 227:1408–1419

Kushmerick C, Romano-Silva MA, Gomez MV, Prado MA (2001) Changes in $Ca^{(2+)}$ channel expression upon differentiation of SN56 cholinergic cells. Brain Res 916:199–210

Lazarovici P, Jiang H, Fink D Jr (1998) The 38-amino-acid form of pituitary adenylate cyclase-activating polypeptide induces neurite outgrowth in PC12 cells that is dependent on protein kinase C and extracellular signal-regulated kinase but not on protein kinase A, nerve growth factor receptor tyrosine kinase, p21(ras) G protein, and pp60(c-src) cytoplasmic tyrosine kinase. Mol Pharmacol 54:547–558

Le Douarin NM (1988) On the origin of pancreatic endocrine cells. Cell 53:169–171

Lewis DL, Goodman MB, St John PA, Barker JL (1988) Calcium currents and fura-2 signals in fluorescence-activated cell sorted lactotrophs and somatotrophs of rat anterior pituitary. Endocrinology 123:611–621

Lichvarova L, Jaskova K, Lacinova L (2012) NGF-induced neurite outgrowth in PC12 cells is independent of calcium entry through L-type calcium channels. Gen Physiol Biophys 31:473–478

Lory P, Bidaud I, Chemin J (2006) T-type calcium channels in differentiation and proliferation. Cell Calcium 40:135–146

Louhivuori LM, Louhivuori V, Wigren HK, Hakala E, Jansson LC, Nordstrom T, Castren ML, Akerman KE (2013) Role of low voltage activated calcium channels in neuritogenesis and active migration of embryonic neural progenitor cells. Stem Cells Dev 22:1206–1219

Lopez-Dominguez AM, Espinosa JL, Navarrete A, Avila G, Cota G (2006) Nerve growth factor affects Ca2+ currents via the p75 receptor to enhance prolactin mRNA levels in GH3 rat pituitary cells. J Physiol 574:349–365

Lukyanetz EA (1998) Diversity and properties of calcium channel types in NG108-15 hybrid cells. Neuroscience 87:265–274

Manecka DL, Mahmood SF, Grumolato L, Lihrmann I, Anouar Y (2013) Pituitary adenylate cyclase-activating polypeptide (PACAP) promotes both survival and neuritogenesis in PC12 cells through activation of nuclear factor kappa B (NF-kappaB) pathway: involvement of extracellular signal-regulated kinase (ERK), calcium and c-REL. J Biol Chem 288:14936–14948

Mansvelder HD, Kits KS (2000) All classes of calcium channel couple with equal efficiency to exocytosis in rat melanotropes, inducing linear stimulus-secretion coupling. J Physiol 526:327–339

Mariot P, Vanoverberghe K, Lalevee N, Rossier MF, Prevarskaya N (2002) Overexpression of an alpha 1H (Cav3.2) T-type calcium channel during neuroendocrine differentiation of human prostate cancer cells. J Biol Chem 277:10824–10833

Monet M, Lehen'kyi V, Gackiere F, Firlej V, Vandenberghe M, Roudbaraki M, Gkika D, Pourtier A, Bidaux G, Slomianny C, Delcourt P, Rassendren F, Bergerat JP, Ceraline J, Cabon F, Humez S, Prevarskaya N (2010) Role of cationic channel TRPV2 in promoting prostate cancer migration and progression to androgen resistance. Cancer Res 70:1225–1235

Mori S, Murakami-Mori K, Bonavida B (1999) Interleukin-6 induces G1 arrest through induction of p27(Kip1), a cyclin-dependent kinase inhibitor, and neuron-like morphology in LNCaP prostate tumor cells. Biochem Biophys Res Commun 257:609–614

Mudado MA, Rodrigues AL, Prado VF, Beirao PS, Cruz JS (2004) $Ca_V3.1$ and $Ca_V3.3$ account for T-type Ca^{2+} current in GH3 cells. Braz J Med Biol Res 37:929–935

Nagasawa K, Tarui T, Yoshida S, Sekiguchi F, Matsunami M, Ohi A, Fukami K, Ichida S, Nishikawa H, Kawabata A (2009) Hydrogen sulfide evokes neurite outgrowth and expression of high-voltage-activated Ca^{2+} currents in NG108-15 cells: involvement of T-type Ca^{2+} channels. J Neurochem 108:676–684

Nakashima S, Ikeno Y, Yokoyama T, Kuwana M, Bolchi A, Ottonello S, Kitamoto K, Arioka M (2003) Secretory phospholipases A2 induce neurite outgrowth in PC12 cells. Biochem J 376:655–666

Novara M, Baldelli P, Cavallari D, Carabelli V, Giancippoli A, Carbone E (2004) Exposure to cAMP and beta-adrenergic stimulation recruits Ca(V)3 T-type channels in rat chromaffin cells through Epac cAMP-receptor proteins. J Physiol 558:433–449

Oesterling JE, Hauzeur CG, Farrow GM (1992) Small cell anaplastic carcinoma of the prostate: a clinical, pathological and immunohistological study of 27 patients. J Urol 147:804–807

Panner A, Wurster RD (2006) T-type calcium channels and tumor proliferation. Cell Calcium 40:253–259

Pearse AG (1968) Common cytochemical and ultrastructural characteristics of cells producing polypeptide hormones (the APUD series) and their relevance to thyroid and ultimobranchial C cells and calcitonin. Proc R Soc Lond B Biol Sci 170:71–80

Pretl K (1944) Zur Frage der Endokrinie der menschlichen Vorsteherdruse. Virchows Arch 312:392–404

Ravni A, Bourgault S, Lebon A, Chan P, Galas L, Fournier A, Vaudry H, Gonzalez B, Eiden LE, Vaudry D (2006) The neurotrophic effects of PACAP in PC12 cells: control by multiple transduction pathways. J Neurochem 98:321–329

Roussel JP, Mateu G, Astier H (1992) Blockade of potassium or calcium channels provokes modifications in TRH-induced TSH release from rat perifused pituitaries. Endocr Regul 26:163–170

Sagnak L, Topaloglu H, Ozok U, Ersoy H (2011) Prognostic significance of neuroendocrine differentiation in prostate adenocarcinoma. Clin Genitourin Cancer 9:73–80

Sainz RM, Mayo JC, Tan DX, Leon J, Manchester L, Reiter RJ (2005) Melatonin reduces prostate cancer cell growth leading to neuroendocrine differentiation via a receptor and PKA independent mechanism. Prostate 63:29–43

Seidah NG (2011) What lies ahead for the proprotein convertases? Ann N Y Acad Sci 1220:149–161

Shen R, Dorai T, Szaboles M, Katz AE, Olsson CA, Buttyan R (1997) Transdifferentiation of cultured human prostate cancer cells to a neuroendocrine cell phenotype in a hormone-depleted medium. Urol Oncol 3:67–75

Sherwood NT, Lesser SS, Lo DC (1997) Neurotrophin regulation of ionic currents and cell size depends on cell context. Proc Natl Acad Sci USA 94:5917–5922

Shitaka Y, Matsuki N, Saito H, Katsuki H (1996) Basic fibroblast growth factor increases functional L-type Ca2+ channels in fetal rat hippocampal neurons: implications for neurite morphogenesis in vitro. J Neurosci 16:6476–6489

Silver RA, Bolsover SR (1991) Expression of T-type calcium current precedes neurite extension in neuroblastoma cells. J Physiol Paris 85:79–83

Tarui T, Fukami K, Nagasawa K, Yoshida S, Sekiguchi F, Kawabata A (2010) Involvement of Src kinase in T-type calcium channel-dependent neuronal differentiation of NG108-15 cells by hydrogen sulfide. J Neurochem 114:512–519

Tojima T, Yamane Y, Takahashi M, Ito E (2000) Acquisition of neuronal proteins during differentiation of NG108-15 cells. Neurosci Res 37:153–161

Vanden Abeele F, Shuba Y, Roudbaraki M, Lemonnier L, Vanoverberghe K, Mariot P, Skryma R, Prevarskaya N (2003) Store-operated Ca2+ channels in prostate cancer epithelial cells: function, regulation, and role in carcinogenesis. Cell Calcium 33:357–373

Vanoverberghe K, Lehen'kyi V, Thebault S, Raphael M, Vanden Abeele F, Slomianny C, Mariot P, Prevarskaya N (2012) Cytoskeleton reorganization as an alternative mechanism of store-operated calcium entry control in neuroendocrine-differentiated cells. PLoS One 7: e45615

Vanoverberghe K, Vanden Abeele F, Mariot P, Lepage G, Roudbaraki M, Bonnal JL, Mauroy B, Shuba Y, Skryma R, Prevarskaya N (2004) Ca^{2+} homeostasis and apoptotic resistance of neuroendocrine-differentiated prostate cancer cells. Cell Death Differ 11:321–330

Weiss N, Hameed S, Fernandez-Fernandez JM, Fablet K, Karmazinova M, Poillot C, Proft J, Chen L, Bidaud I, Monteil A, Huc-Brandt S, Lacinova L, Lory P, Zamponi GW, De Waard M (2012) A Ca(v)3.2/syntaxin-1A signaling complex controls T-type channel activity and low-threshold exocytosis. J Biol Chem 287:2810–2818

Weiss N, Zamponi GW (2013) Control of low-threshold exocytosis by T-type calcium channels. Biochim Biophys Acta 1828:1579–1586

Westerink RH, Ewing AG (2008) The PC12 cell as model for neurosecretion. Acta Physiol (Oxf) 192:273–285

Wu C, Huang J (2007) Phosphatidylinositol 3-kinase-AKT-mammalian target of rapamycin pathway is essential for neuroendocrine differentiation of prostate cancer. J Biol Chem 282:3571–3583

Yao JL, Madeb R, Bourne P, Lei J, Yang X, Tickoo S, Liu Z, Tan D, Cheng L, Hatem F, Huang J, Anthony di Sant'Agnese P (2006) Small cell carcinoma of the prostate: an immunohistochemical study. Am J Surg Pathol 30:705–712

Yuan TC, Veeramani S, Lin FF, Kondrikou D, Zelivianski S, Igawa T, Karan D, Batra SK, Lin MF (2006) Androgen deprivation induces human prostate epithelial neuroendocrine differentiation of androgen-sensitive LNCaP cells. Endocr Relat Cancer 13:151–167

Yuan TC, Veeramani S, Lin MF (2007) Neuroendocrine-like prostate cancer cells: neuroendocrine transdifferentiation of prostate adenocarcinoma cells. Endocr Relat Cancer 14:531–547

Zhang XQ, Kondrikov D, Yuan TC, Lin FF, Hansen J, Lin MF (2003) Receptor protein tyrosine phosphatase alpha signaling is involved in androgen depletion-induced neuroendocrine differentiation of androgen-sensitive LNCaP human prostate cancer cells. Oncogene 22:6704–6716

Regulation of Cardiac Hypertrophy by T-Type Ca^{2+} Channel

5

Stephen W. Schaffer and Chian Ju Jong

Abstract

Congestive heart failure is a terminal condition whose progression is determined by both cardiomyocyte and ventricular remodeling. Among the pathways contributing to remodeling are the calcium-dependent calcineurin–NFAT and CREB hypertrophic pathways. While the T-type Ca^{2+} channel and the transient receptor potential anion channel superfamily assume central roles in initiation of cardiac hypertrophy involving the calcineurin–NFAT pathway, the T-type Ca^{2+} channel appears to regulate the calcineurin–NFAT hypertrophic pathway. The T-type Ca^{2+} channel also modulates the development of pathological hypertrophy via the CREB hypertrophic pathway.

5.1 Physiological Roles of Myocardial Ca^{2+} Channels

The recognition that stimulation of Purkinje fibers leads to a rapid increase in membrane potential associated with the influx of Na^+ ions followed by a plateau in membrane potential, a phase dominated by the slow influx of Ca^{2+} ions (referred to as I_{si} or slow inward current), was an important breakthrough in the area of excitation–contraction coupling (Reuter 1967). It is now known that in normal adult ventricular cardiomyocyte, I_{si} is synonymous with $I_{Ca,L}$, the large and long-lasting Ca^{2+} current associated with activation of L-type Ca^{2+} channels (Nowycky et al. 1985). Hence, the L-type Ca^{2+} channel is largely responsible for filling the sarcoplasmic reticular Ca^{2+} stores and triggering Ca^{2+}-induced Ca^{2+} release from the sarcoplasmic reticulum, key events in excitation–contraction (EC) coupling. Because of species differences in the structure of the sarcoplasmic reticulum, the

S.W. Schaffer (✉) • C.J. Jong
Department of Pharmacology, University of South Alabama College of Medicine, Mobile, AL 36688, USA
e-mail: sschaffe@southalabama.edu

S.W. Schaffer, M. Li (eds.), *T-type Calcium Channels in Basic and Clinical Science*,
DOI 10.1007/978-3-7091-1413-1_5

importance of sarcoplasmic reticular Ca^{2+} handling differs, with the frog heart largely lacking the sarcoplasmic reticular Ca^{2+} release mechanism while the rat heart being highly dependent on Ca^{2+}-induced Ca^{2+} release for contractile function (Bers 2001). The L-type Ca^{2+} channel, which is encoded by two different genes ($Ca_v1.2$ and $Ca_v1.3$) in the heart, is also required for normal heart development, as inactivation of the $Ca_v1.2$ gene leads to embryonic lethality (Xu et al. 2003). However, elevations in $I_{Ca,L}$ flux by overexpression of the L-type Ca^{2+} channel lead to Ca^{2+} overload, which in turn produces a phenotype with reduced contractile function and elevated cellular necrosis (Jaleel et al. 2008).

Atrial myocytes, Purkinje cells, embryonic and fetal cardiomyocytes, and diseased cardiomyocytes contain a second type of voltage-dependent Ca^{2+} channel, termed the T-type Ca^{2+} channel. The expression of the T-type Ca^{2+} channel is virtually absent in most adult cardiomyocytes, although the guinea pig heart contains significant levels of the T-type Ca^{2+} channel and the T-type Ca^{2+} channel is reexpressed in the hypertrophied heart, suggesting that the channel is part of the fetal gene program that emerges during the hypertrophic process. The T-type Ca^{2+} channel derived its name from two characteristic features of the current, "tiny and transient." Because $I_{Ca,T}$ is both tiny and transient by virtue of its rapid inactivation, it is incapable of being a major trigger of sarcoplasmic reticular Ca^{2+} release in the heart. Moreover, $I_{Ca,T}$ is less efficient at triggering Ca^{2+}-induced Ca^{2+} release than $I_{Ca,L}$ because the T-type Ca^{2+} channels are located on the surface of the sarcolemma while the L-type Ca^{2+} channels are found within the T-tubules in close proximity to the ryanodine release channels (Sipido et al. 1998; Jaleel et al. 2008). Consequently, T-type Ca^{2+} channel-dependent EC coupling is less effective, resulting in delayed onset of cell shortening, slower rates of shortening and relaxation, lower peak shortening, and longer time to peak shortening relative to L-type Ca^{2+} channel-dependent EC coupling (Zhou and January 1998). However, several lines of evidence support at least some role for T-type Ca^{2+} channels in EC coupling. First, Sipido et al. (1998) showed that Ca^{2+} release was greater in guinea pig ventricular myocytes in the presence of both $I_{Ca,T}$ and $I_{Ca,L}$ than at holding voltages allowing only $I_{Ca,L}$. Second, based on the inhibitory actions of nifedipine, Kitchens et al. (2003) concluded that $I_{Ca,T}$ is an important source of trigger Ca^{2+} in chick embryo cardiomyocytes. Indeed, the amplitude of Ca^{2+} transients is more dependent on the T-type Ca^{2+} channel in the neonatal rat heart while the L-type Ca^{2+} channel assumes a central role in Ca^{2+} cycling in the adult heart (Escobar et al. 2004). Third, T-type Ca^{2+} channel antagonists depress contractile function in guinea pig cardiac myocytes, which express both T-type and L-type Ca^{2+} channels (Hoischen et al. 1998). Fourth, overexpression of the T-type Ca^{2+} channel gene $Ca_v3.1$, which significantly elevates $I_{Ca,T}$, also increases the amplitude of the Ca^{2+} transient and contractile performance (Nakayama et al. 2009).

Two distinct genes, $Ca_v3.1$ (α1G) and $Ca_v3.2$ (α1H), code for T-type Ca^{2+} channels in myocardial ventricles (Cribbs et al. 1998; Monteil et al. 2000). Based on its hydropathy profile, the $Ca_v3.1$ subunit of the T-type Ca^{2+} channel consists of 24 transmembrane segments arranged in four domains of six transmembrane segments each (Talavera and Nilius 2006). The intracellular part of the channel is formed by segment 6 while the P loops make up the extracellular section of the channel. Segment 4 of each transmembrane domain is positively charged and confers sensitivity to membrane potential changes while Ca^{2+} selectivity and activation kinetics are determined by two glutamate residues found in domains I and II and two aspartate residues found in domains III and IV (Talavera et al. 2001; Perez-Reyes 2003; Talavera and Nilius 2006). Overexpression of $Ca_v3.1$ in mice elevates Ca^{2+} influx, sarcoplasmic reticular Ca^{2+} loading, and contractile function, implying that stimulation of $I_{Ca,T}$ increases the availability of Ca^{2+} for contraction. However, the change in $Ca_v3.1$ content is associated with changes in the content of other Ca^{2+} regulatory proteins, including the L-type Ca^{2+} channel, SERCA2a, and phospholamban. Thus, the improvement in contractile function in the $Ca_v3.1$-overloaded heart could be attributed to changes in one or more of the other Ca^{2+} regulatory proteins rather than an increase in $I_{Ca,T}$.

5.2 T-Type Ca^{2+} Channel Involvement in Myocardial Hypertrophy and Heart Failure

Heart failure is defined as the final clinical presentation of a variety of cardiovascular diseases, ranging from coronary artery disease, hypertension, valvular heart disease, myocarditis, and alcoholic cardiomyopathy. The initial event in each of these cardiovascular diseases is a reduction in cardiac output. In response to impaired contractile function, the heart undergoes an adaptation that initially increases cardiac size and improves contractile function. Key events involved in the adaptation process are elevations in the blood levels of norepinephrine and members of the renin–angiotensin II–aldosterone system. These neurohumoral factors initiate a complex series of events that lead to ventricular remodeling and cardiomyocyte death; the development of cardiac hypertrophy is an important facet of ventricular remodeling (Scheuer 1999). The development of both receptor-stimulated and pressure overload-induced pathological hypertrophy involves angiotensin II and norepinephrine signaling. Pathological hypertrophy contrasts with physiological hypertrophy, which is mediated by growth factors, volume loading, and aerobic exercise training that proceeds through Akt and mammalian target of rapamycin (mTOR) (Kurdi and Booz 2011).

5.2.1 Initiation of Hypertrophy by Angiotensin II and Norepinephrine

Molkentin et al. (1998) and Onohara et al. (2006) proposed a signaling pathway (calcineurin–NFAT pathway) for the development of myocardial hypertrophy initiated by the $G_{\alpha q}$ protein-coupled receptors, a group that includes the receptors for both norepinephrine (α_1-adrenergic receptor) and angiotensin II (AT_1 receptor) (Fig. 5.1). Activation of the G protein-coupled receptors leads to the activation of phospholipase C, an enzyme that catalyzes the generation of diacylglycerol and 1,4,5-triphosphate (IP_3). IP_3 interacts with its receptor to elevate $[Ca^{2+}]_i$ while

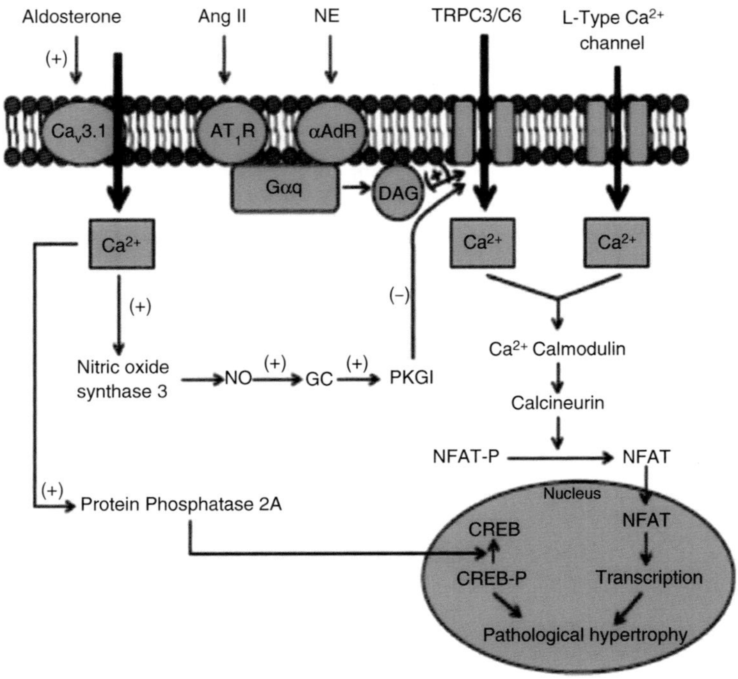

Fig. 5.1 Scheme of the calcineurin–NFAT and CREB hypertrophic pathways. The calcineurin–NFAT hypertrophic pathway is initiated by the interaction of angiotensin II (Ang II) and norepinephrine with their respective $G_{\alpha q}$ receptors, which leads to the generation of diacylglycerol (DAG), a stimulator of the TRPC (transient receptor potential C) family. Activation of the TRPC3/TRPC6 dimer complex is coupled to the activation of the L-type Ca^{2+} channel, leading to enhanced Ca^{2+} influx. The resulting increase in $[Ca^{2+}]_i$ leads to the formation of Ca^{2-} calmodulin, which in turn activates the protein phosphatase calcineurin. Calcineurin catalyzes the dephosphorylation of NFAT (nuclear factor of activated T cells), which is translocated to the nucleus where it promotes transcription, a step in the process of pathological hypertrophy. The calcineurin–NFAT pathway is inhibited by a pathway initiated by aldosterone-mediated upregulation of $Ca_v3.1$. As a component of the T-type Ca^{2+} channel, $Ca_v3.1$ increases the Ca^{2+} pool that activates nitric oxide synthase 3, an enzyme that catalyzes the formation of nitric oxide (NO). Nitric oxide activates guanylate cyclase (GC), which catalyzes the formation of cGMP, which in turn activates PKGI (cGMP-dependent protein kinase I). The protein kinase mediates an anti-hypertrophic response by inhibiting the TRPC3/TRPC6 complex. $Ca_v3.1$ also promotes an alternate anti-hypertrophic pathway involving Ca^{2+}-dependent activation of protein phosphatase 2A. The phosphatase catalyzes the dephosphorylation of CREB (cAMP-responsible element binding protein), which diminishes transcription and the development of pathological hypertrophy

diacylglycerol increases $[Ca^{2+}]_i$ by stimulation of TRPC3 and TRPC6, members of the transient receptor potential (TRP) cation channel superfamily (Nilius et al. 2007; Eder and Mokentin 2011). The increase in $[Ca^{2+}]_i$ occurs upon stimulation of the L-type Ca^{2+} channel by the TRPC proteins (Onohara et al. 2006). The calcineurin–NFAT arm of the pathway lies downstream from the TRPC and L-type Ca^{2+} channels (Onohara et al. 2006). Although activation of the TRPC proteins is required for angiotensin II-mediated cardiomyocyte hypertrophy, other sources of cytosolic Ca^{2+} may contribute to the hypertrophic process. Irrespective of the source of Ca^{2+}, the rise in $[Ca^{2+}]_i$ increases the amount of Ca^{2+} bound to calmodulin in the cytosol. In its Ca^{2+}–calmodulin form, calmodulin can stimulate the protein phosphatase calcineurin. Upon activation, calcineurin catalyzes the dephosphorylation of NFAT (nuclear factor of activated T cells), which is then translocated to the nucleus where it enhances protein synthesis. Because NFAT3 null mice are resistant to hypertrophic development, it has been proposed that stimulation of the calcineurin–NFAT pathway is a key event in the development of myocardial hypertrophy (Wilkins et al. 2002).

The TRPC family of cation channels has attracted considerable interest because the activation of TRPC3 and TRPC6 is required for angiotensin II-induced NFAT activation and subsequent development of pathological hypertrophy (Onohara et al. 2006). However, many other members of the TRP family (TRPC1, TRPC4, TRPC5, TRPC7, TRPV2, and TRPM4) have been implicated in myocardial hypertrophy and the pathogenesis of heart failure (Nishida and Kurose 2008). TRPC1 appears to be a component of stretch-activated cation channels, which are involved in stretch-induced cardiomyocyte hypertrophy (Maroto et al. 2005; Ohba et al. 2007). Downregulation of TRPC1 renders mouse hearts resistant to pressure overload- and angiotensin II-induced cardiac hypertrophy and contractile dysfunction, effects associated with diminished calcineurin–NFAT signaling (Seth et al. 2009). It has also been suggested that TRPC1 might contribute to the leakage of Ca^{2+} from the sarcoplasmic reticulum (Berbey et al. 2009). TRPC3 is upregulated by activation of Ca^{2+}-sensing receptors (Feng et al. 2011). Bush et al. (2006) found that overexpression of TRPC3 in the neonatal cardiomyocyte increases cell size and promotes the upregulation of fetal genes through stimulation of calcineurin–NFAT signaling. Poteser et al. (2011) showed that Ca^{2+} permeation through the TRPC3 channel is required for the activation of the calcineurin–NFAT pathway. The coupling between TRPC3 and the calcineurin–NFAT pathway is dependent on PKC-mediated regulation of TRPC3 (Poteser et al. 2011). It has also been proposed that overexpression of TRPC3 leads to sustained elevations in $[Ca^{2+}]_i$ derived from store-operated Ca^{2+} channels. The excessive increase in $[Ca^{2+}]_i$ causes the death of some cardiomyocytes via apoptosis (Feng et al. 2011) and the development of a cardiomyopathy characterized by elevations in store-operated Ca^{2+} channels and cardiac hypertrophy (Nakayama et al. 2006). Like TRPC1, both TRPC4 and TRPC5 are associated with store-operated Ca^{2+} current that is regulated by IP_3 receptors (Venkatachalam et al. 2003; Eder and Mokentin 2011). Downregulation of TRPC4 diminishes Ca^{2+} influx and the development of pathological cardiac hypertrophy while overexpression of TRPC4 increases β_1-adrenergic receptor-mediated activation of NFAT (Wu et al. 2010). Interestingly, TRPC3 and TRPC4 co-assemble to generate a redox-sensitive cation channel in endothelial cells (Poteser et al. 2006). Inhibition of both TRPC3 and TRPC4 completely abolish transverse aortic constriction-associated Ca^{2+} entry (Wu et al. 2010). TRPC5 mRNA and protein expression are elevated in the human failing heart (Bush et al. 2006). It has been suggested that TRPC5 plays a role in dilated cardiomyopathy rather than the development of cardiac hypertrophy (Nishida and Kurose 2008). TRPC6 is an inwardly and outwardly rectifying channel that exhibits low basal activity. TRPC6 shares considerable homology with TRPC3 and TRPC7, accounting for similar modes of regulation by diacylglycerol and protein kinase C (Venkatachalam et al. 2003). All three family members can associate to form heterotrimer or heterodimer complexes, with the TRPC3/6 channel implicated in angiotensin II-mediated elevations in Ca^{2+} influx and subsequent activation of the calcineurin–NFAT signaling pathway (Onohara et al. 2006). Overexpression of TRPC6 has been shown to increase cation current, stimulate the translocation of NFAT from the cytosol to the nucleus, and cause cardiac hypertrophy (Kuwahara et al. 2006; Onohara et al. 2006). Among the genes implicated in the actions of NFAT are BNP, β-MHC, RCAN, and TRPC6, with β-MHC expression acting to reduce contractile function, BNP representing a return to myocardial fetal expression, and both RCAN and TRPC6 acting to amplify calcineurin–NFAT signaling (Kuwahara et al. 2006). Kuwahara et al. (2006) found that robust TRPC6 overexpression leads to the development of a severe cardiomyopathy that can be fatal while mild overexpression of TRPC6 does not directly lead to cardiac hypertrophy but does sensitize the heart to the effects of pressure overload. Satoh et al. (2007) reported that overexpression of TRPC7 leads to apoptosis although their conclusions have been challenged by the work of Nishida and Kurose (2008). The major regulators of the TRPC family in the heart are the neurohumoral factors and pressure overload (Richard et al. 1998; Muth et al. 1999; Brette et al. 2006; Nakayama et al. 2006; Wu et al. 2010; Chen et al. 2011; Gao et al. 2012; Makarewich et al. 2012). Based on mRNA analyses, the expression profile of the TRPC family is TRPC1 > TRPC4 > TRPC6 > TRPC5 ≫ TRPC7 (Friedrich et al. 2012).

5.3 Potential Role of T-Type Ca^{2+} Channels in Development of Pathological Hypertrophy

Because the calcineurin–NFAT pathway contributes to the development of pathological hypertrophy, the source of Ca^{2+} for calcineurin activation has been an area of considerable interest and debate. One of the observations that has dramatically altered the perception of intracellular Ca^{2+} as a second messenger was the realization that intracellular Ca^{2+} exists as distinct Ca^{2+} pools and that each specialized pool serves as a regulator of a specific signaling protein, including calcineurin, protein kinase C, and Ca^{2+} calmodulin-dependent protein kinase II (CaMKII) (Balijepalli et al. 2006; Wu et al. 2006). In agreement with this notion, Jaleel et al. (2008) advanced the view that the local Ca^{2+} pool fed by the T-type Ca^{2+} channel differs from that of the L-type Ca^{2+} channel, raising the possibility that the two Ca^{2+} pools must serve different functions. While the L-type Ca^{2+} channel plays a central role in excitation–contraction coupling and in the activation of the calcineurin–NFAT pathway, the T-type Ca^{2+} channel must serve another function. This concept assumes that the L-type Ca^{2+} channel and members of the TRP (transient receptor potential) channels are coupled, a viewpoint supported by data showing that (1) nifedipine not only inhibits the L-type Ca^{2+} channel but also TRP channel 3-induced activation of the calcineurin–NFAT pathway, (2) reduced levels of TRP channel 3 are associated with diminished L-type Ca^{2+} current and calcineurin–NFAT signaling, and (3) both nitrendipine (L-type Ca^{2+} channel antagonist) and SK&F96365 (receptor activated cation channel antagonist) block angiotensin II-mediated NFAT activation (Onohara et al. 2006; Gao et al. 2012).

Although the T-type Ca^{2+} channel is not directly associated with the local Ca^{2+} pool involved in the activation of the calcineurin–NFAT pathway, some investigators feel that the T-type Ca^{2+} channel might contribute somewhat to the development of cardiac hypertrophy (Takebayshi et al. 2006; Horiba et al. 2008; Chiang et al. 2009), an idea propelled by the observation that T-type Ca^{2+} channels are reexpressed in both hypertrophied hearts and cardiomyocytes treated with hypertrophic stimulants (Martinez et al. 1999; Huang et al. 2000; Ferron et al. 2003; Izumi et al. 2003; Bkaily et al. 2005; Morishima et al. 2009). Takebayshi et al. (2006) also reported an increase in $Ca_v3.1$ and $Ca_v3.2$ mRNA in hypertrophied right ventricles of monocrotaline-treated hearts.

Five lines of evidence support a role for the T-type Ca^{2+} channel in the development of pathological hypertrophy. First, Chiang et al. (2009) found that $Ca_v3.2^{-/-}$, but not $Ca_v3.1^{-/-}$, mice are resistant to pressure overload-induced hypertrophy. The possibility that $Ca_v3.2$ might contribute to the Ca^{2+} pool regulating calcineurin activity was proposed. Second, neonatal cardiomyocytes isolated from $Ca_v3.2^{-/-}$ mice do not develop hypertrophy in response to angiotensin II. Third, NFAT–luciferase reported activity does not respond to pressure overload in the $Ca_v3.2^{-/-}$ mouse, suggesting that $Ca_v3.2$ is involved in the development of hypertrophy through the calcineurin–NFAT hypertrophic pathway. Fourth, T-type Ca^{2+} channel inhibitors (kurtoxin and mibefradil) reduce the degree of ventricular remodeling more than nifedipine (Takebayshi et al. 2006; Horiba et al. 2008; Jaleel et al. 2008) although this finding is inconsistent with a study by Gao et al. (2012), who found that 10 μM nifedipine is more effective than 50 μM Ni^{2+} in preventing either Ca^{2+}-induced or TRP-mediated NFAT nuclear translocation and cardiomyocyte hypertrophy. Fifth, treatment with T-type Ca^{2+} channel antagonists has been shown to limit myocardial damage in several animal models of heart failure (Sandmann et al. 1998, 1999, 2001; Villame et al. 2001).

Human studies examining the effect of T-type Ca^{2+} channel antagonists on the development of heart failure have been equivocal. Hoglund et al. (1998) found that mibefradil (50–100 mg) treatment diminishes left ventricular hypertrophy and blood pressure in a group of 66 mild-to-moderate hypertensive patients. On the other hand, mibefradil (50–100 mg/day) therapy exerts no apparent benefit relative to morbidity and mortality in the MACH-1 study of 2,590 patients suffering from moderate-to-severe congestive heart failure (Levine et al. 2000). However, these studies, while clinically relevant, provide limited information on the effect of the T-type Ca^{2+} channels, as mibefradil is no longer considered a specific T-type Ca^{2+}

channel antagonist and is associated with both drug–drug interactions and untoward side effects.

In contrast to evidence supporting a role for the T-type Ca^{2+} channel in the development of pathological hypertrophy, there are several studies concluding that the T-type Ca^{2+} channel may antagonize cardiac hypertrophy rather than promote it. According to Nakayama et al. (2009), the Ca^{2+} pool fed by $Ca_v3.1$ induces nitric oxide synthase 3, whose activation generates cGMP and promotes cGMP-dependent protein kinase I (PKGI), which mediates an anti-hypertrophic response (Fig. 5.1) (Cappola et al. 2003; Ichinose et al. 2004). In their study, overexpression of $Ca_v3.1$ significantly increased Ca^{2+} cycling and elevated contractile function, but the transgenic mice were resistant to cardiac hypertrophy mediated by pressure overload (transverse aortic constriction), isoproterenol infusion, and exercise. Moreover, $Ca_v3.1^{-/-}$ mice showed enhanced susceptibility to hypertrophic stimulation (isoproterenol infusion and swimming exercise), an effect reversed following restoration of $Ca_v3.1$ content. Based on the finding that nitric oxide synthase 3 coprecipitates with $Ca_v3.1$ after pressure overload and that PKGI activity is elevated more in $Ca_v3.1$-overexpressed hearts than control hearts following pressure overload, it was concluded that $Ca_v3.1$ provides Ca^{2+} for a local pool that activates nitric oxide synthase 3 to mediate an anti-hypertrophic effect through cGMP and PKGI (Nakayama et al. 2009). In further support of the anti-hypertrophic hypothesis, Kinoshita et al. (2010) found that the anti-hypertrophic effect of guanylyl cyclase A can be traced to the inhibition of TRPC6. Moreover, cardiac atrial natriuretic peptide mediates its anti-hypertrophic actions through PKGI and its target, regulator of G protein signaling 2 (Klaiber et al. 2010).

Another mechanism advanced to explain the antihypertensive activity of the T-type Ca^{2+} channel maintains that aldosterone-mediated induction of $Ca_v3.1$ increases $I_{Ca,T}$ (Ferron et al. 2011). The elevation in the size of a T-type Ca^{2+} channel-dependent intracellular Ca^{2+} pool results in the activation of protein phosphatase 2A and to a lesser extent protein phosphatase 1. The activation of protein phosphatase 2A promotes the dephosphorylation of CREB (cAMP-responsive element binding protein), thereby inactivating the transcription factor and limiting the degree of cardiomyocyte hypertrophy.

Conclusions

The mammalian heart contains two major voltage-dependent Ca^{2+} channels, the L-type and the T-type. $I_{Ca,L}$ is relatively large and of fairly long duration, in contrast to the $I_{Ca,T}$, which is transient and tiny. They also are located in different regions of the cell, which has led to the notion that they must serve different physiological functions. The L-type Ca^{2+} channel is recognized for its central role in excitation–contraction coupling and in the development of pathological hypertrophy. It has been proposed that the T-type Ca^{2+} channel contributes to excitation–contraction coupling in the immature heart, but in the adult heart its contribution is limited even when overexpressed. Using similar logic, the T-type Ca^{2+} channel likely plays a minor role in the development of pathological hypertrophy, with the L-type Ca^{2+} channel and the TRPC proteins serving as the major sources of Ca^{2+} for the activation of the calcineurin–NFAT pathway. Therefore, two key questions remain to be answered: (1) what is the physiological function of the T-type Ca^{2+} channel in the heart and (2) does the T-type Ca^{2+} channel alter the condition of the diseased heart? The present review discusses a novel hypothesis for the T-type Ca^{2+} channel, implicating it in the regulation of hypertrophy. According to one scheme, the T-type Ca^{2+} channel provides Ca^{2+} for the activation of nitric oxide synthase 3, which in turn exerts anti-hypertrophic actions through inhibition of the TRPC proteins. In the other scheme, the anti-hypertrophic effect of the T-type Ca^{2+} channel involves Ca^{2+}-dependent activation of protein phosphatase 2A, which dephosphorylates CREB to disrupt its pro-growth properties. Although the concept that the T-type Ca^{2+} channel provides Ca^{2+} to a pool that exerts anti-hypertrophic activity is controversial, new studies examining the actions of the T-type Ca^{2+} channel on the hypertrophic process are warranted. Moreover, the possibility that $Ca_v3.1$ and $Ca_v3.2$ might exert opposite actions on the hypertrophic process deserves further consideration.

References

Balijepalli RC, Foell JD, Hall DD, Hell JW, Kamp TJ (2006) Localization of cardiac L-type Ca^{2+} channels to a caveolar macromolecular signaling complex is required for β_2–adrenergic regulation. Proc Natl Acad Sci USA 103:7500–5

Berbey C, Weiss N, Legrand C, Allard B (2009) Transient receptor potential canonical type 1 (TRPC1) operates as a sarcoplasmic reticulum calcium leak channel in skeletal muscle. J Biol Chem 284:36387–94

Bers DM (2001) Excitation-contraction coupling and cardiac contractile force. Kluwer, Dordrecht, Netherlands

Bkaily G, Sculptoreanu A, Wang S, Nader M, Hazzouri KM, Jacques D, Regoli D, D'Orleans-Juste P, Avendanian L (2005) Angiotensin II-induced increase of T-type Ca^{2+} current and decrease of L-type Ca^{2+} current in heart cells. Peptides 26:1410–7

Brette F, Leroy J, Le Guennec J-Y, Salle L (2006) Ca^{2+} currents in cardiac myocytes: old story, new insight. Prog Biophys Mol Biol 91:1–82

Bush EW, Hood DB, Papst PJ, Chapo JA, Minobe W, Bristow MR, Olson EN, McKinsey TA (2006) Canonical transient receptor potential channels promote cardiomyocyte hypertrophy through activation of calcineurin signaling. J Biol Chem 281:33487–96

Cappola TP, Cope L, Cernetich A, Barouch LA, Minhas K, Irizarry RA, Parmigiani G, Durrani S, Lavoie T, Hoffman EP, Ye SQ, Garcia JG, Hare JM (2003) Deficiency of different nitric oxide synthase isoforms activates divergent transcriptional programs in cardiac hypertrophy. Physiol Genomics 14:25–34

Chen X, Nakayama H, Zhang X, Ai X, Harris DM, Tang M, Zhang H, Szeto C, Stockbower K, Berretta RM, Eckhart AD, Koch WJ, Mokentin JD, Houser SR (2011) Calcium influx through $Ca_v1.2$ is a proximal signal for pathological cardiomyocyte hypertrophy. J Mol Cell Cardiol 50:460–70

Chiang CS, Huang CH, Chieng H, Chang YT, Chang D, Chen JJ, Chen YC, Chen YH, Shin HS, Campbell KP, Chen CC (2009) The Ca_v32. T-type Ca^{2+} channel is required for pressure overload-induced cardiac hypertrophy in mice. Circ Res 104:522–30

Cribbs LL, Lee JH, Ynag J, Satin J, Zhang Y, Daud A, Barclay J, Williamson MP, Fox M, Rees M, Perez-Reyes E (1998) Cloning and characterization of alpha 1H from human heart, a member of the T-type Ca^{2+} channel family. Circ Res 83:103–10

Eder P, Mokentin JD (2011) TRPC channels as effectors of cardiac hypertrophy. Circ Res 108:265–72

Escobar AL, Ribeiro-Costa R, Villalba-Galea C, Zoghbi ME, Perez CG, Mejia-Alvarez R (2004) Developmental changes of intracellular Ca^{2+} transients in beating rat hears. Am J Physiol 286: H971–8

Feng S-L, Sun M-R, T-t L, Yin X, Xu C-Q, Sun Y-H (2011) Activation of calcium-sensing receptor increases TRPC3 expression. Biochem Biophys Res Commun 406:278–84

Ferron L, Capuano V, Ruchon Y, Deroubaix E, Coulombe A, Renaud JF (2003) Signaling pathways mediate expression of cardiac T-type calcium channels. Circ Res 93:1241–8

Ferron L, Ruchon Y, Renaud J-F, Capuano V (2011) T-type Ca^{2+} signaling regulates aldosterone-induced CREB activation and cell death through PP2A activation in neonatal cardiomyocytes. Cardiovasc Res 90:105–12

Friedrich O, Wagner S, Battle AR, Schurmann S, Bartinac B (2012) Mechano-regulation of the beating heart at the cellular level—mechanosensitive channels in normal and diseased heart. Prog Biophys Mol Biol 110:226–38

Gao H, Wang F, Makarewich CA, Zhang H, Kubo H, Berretta RM, Barr LA, Jolkentin JD, Houser SR (2012) Ca(2+) influx through L-type Ca(2+) channels and transient receptor potential channels activates pathological hypertrophy signaling. J Mol Cell Cardiol 53:657–67

Hoglund C, Cifkova R, Mimran A, Tenczer J, Watt A, Wilkins MR, Lindberg E (1998) A comparison of the effects of mibefradil and atenolol on regression of left ventricular hypertrophy in hypertensive patients. Cardiology 89:263–70

Hoischen S, Brixius K, Schwinger RH (1998) T- and L-type Ca^{2+} channel antagonists reduce contractility in guinea pig cardiac myocytes. J Cardiovasc Pharmacol 32:323–330

Horiba M, Muto T, Ueda N, Opthof T, Miwa K, Hojo M, Lee JK, Kamiya K, Kodama I, Yasui K (2008) T-type Ca^{2+} channel blockers prevent cardiac cell hypertrophy through an inhibition of calcineurin-NFAT3 activation as well as L-type Ca^{2+} channel blockers. Life Sci 82:554–60

Huang B, Qin D, Deng L, Boutjdir M, El-Sherif N (2000) Reexpression of T-type Ca^{2+} channel gene and current in post-infarction remodeled rat left ventricle. Cardiovasc Res 46:442–9

Ichinose F, Bloch KD, Wu JC, Hataishi R, Aretz HT, Picard MH, Scherrer-Crosbie M (2004) Pressure overload-induced LV hypertrophy and dysfunction in mice are exacerbated by congenital NOS3 deficiency. Am J Physiol 286:H1070–5

Izumi T, Kihara Y, Sarai N, Yoneda T, Iwanaga Y, Inagaki K, Onozawa Y, Takenaka H, Kita T, Noma A (2003) Reinduction of T-type Ca^{2+} channels by endothelin-1 in failing hearts in vivo and in adult rat ventricular myocytes in vitro. Circulation 108:2530–5

Jaleel N, Nakayama H, Chen X, Kubo H, MacDonnell S, Zhang H, Berretta R, Robbins J, Cribbs L, Mokentin JD, Houser SR (2008) Ca^{2+} influx through T- and L-type Ca^{2+} channels have different effects on myocyte contractility and induce unique cardiac phenotypes. Circ Res 103:1109–19

Kinoshita H, Kuwahara K, Nishida M, Jian Z, Rong X, Kiyonaka S, Kuwabara Y, Kurose H, Inoue R, Mori Y, Li Y, Nakagawa Y, Usami S, Fujiwara M, Yamada Y, Minami T, Ueshima K, Nakao K (2010) Inhibition of TRPC6 channel activity contributes to the antihypertrophic effects of natriuretic peptides-guanylyl cyclase A signaling in the heart. Circ Res 106:1849–60

Kitchens SA, Burch J, Creazzo TL (2003) T-type Ca^{2+} channel current contribution to Ca^{2+}-induced Ca^{2+} release in developing myocardium. J Mol Cell Cardiol 35:515–23

Klaiber M, Kruse M, Volker K, Schroter J, Feil R, Freichel M, Gerling A, Feil S, Dietrich A, Camacho Londono JE, Baba HA, Abramowitz J, Birnbaumer L, Penninger JM, Pongs O, Kuhn M (2010) Novel insights into the mechanisms mediating the local antihypertrophic effects of cardiac atrial natriuretic peptide: role of cGMP-dependent protein kinase and RGS2. Basic Res Cardiol 105:583–95

Kurdi M, Booz GW (2011) Three 4-letter words of hypertension-related cardiac hypertrophy: TRPC, mTOR and HDAC. J Mol Cell Cardiol 50:964–71

Kuwahara K, Wang Y, McAnally J, Richardson JA, Bassel-Duby R, Hill JA, Olson EN (2006) TRPC6 fulfills a calcineurin signaling circuit during pathologic cardiac remodeling. J Clin Invest 116:3114–3126

Levine TB, Beernink PJLM, Caspi A, Elkayam U, Geltman EM, Greenberg B, McKenna WJ, Ghali JK, Giles TD, Marmor A, Reisin LH, Ammon S, Lindberg E (2000) Effect of mibefradil, a T-type Ca^{2+} channel blocker, on morbidity and mortality in moderate to severe congestive heart failure: the MACH-1 study. Circulation 101:758–64

Makarewich CA, Correll RN, Gao H, Zhang H, Yang B, Berretta RM, Rizzo V, Mokentin JD, Houser SR (2012) A caveolae-targeted L-type Ca^{2+} channel antagonist inhibits hypertrophic signaling without reducing cardiac contractility. Circ Res 110:669–74

Maroto R, Raso A, Wood TG, Kurosky A, Martinac B, Hamill OP (2005) TRPC1 forms the stretch-activated cation channel in vertebrates cells. Nat Cell Biol 7:179–85

Martinez ML, Heredia MP, Delgado C (1999) Expression of T-type Ca^{2+} channels in ventricular cells from hypertrophied rat hearts. J Mol Cell Cardiol 31:1617–25

Molkentin JD, Lu JR, Antos CL, Markham B, Richardson J, Robbins J, Grant SR, Olson EN (1998) A calcineurin-dependent transcriptional pathway for cardiac hypertrophy. Cell 93:215–28

Monteil A, Chemin J, Leuranguer V, Altier C, Mennessier G, Bourinet E, Lory P, Nargeot J (2000) Specific properties of T-type calcium channel generated by the human alpha 1I subunit. J Biol Chem 275:16530–5

Morishima M, Wang Y, Akiyoshi Y, Miyamoto S, Ono K (2009) Telmisartan, an angiotensin II type 1 receptor antagonist, attenuates T-type Ca^{2+} channel expression in neonatal rat cardiomyocytes. Eur J Pharmacol 609:105–112

Muth JN, Yamaguchi H, Mikala G, Grupp IL, Lewis W, Cheng H, Song LS, Lakatta EG, Varadi G, Schwartz A (1999) Cardiac-specific overexpression of the alpha 1 subunit of the L-type Ca^{2+} channel in transgenic mice. Loss of isoproterenol-induced contraction. J Biol Chem 274:21503–6

Nakayama H, Wilkin BJ, Bodi I, Mokentin JD (2006) Calcineurin-dependent cardiomyopathy is activated by TRPC in the adult mouse heart. FASEB J 20:1660–70

Nakayama H, Bodi L, Correll RN, Chen X, Lorenz J, Houser SR, Robbins J, Schwartz A, Molkentin JD (2009) Alpha 1G-dependent T-type Ca^{2+} channel current antagonizes cardiac hypertrophy through a NOS3-dependent mechanism in mice. J Clin Invest 119:3787–96

Nilius B, Owsianik G, Voets T, Peters JA (2007) Transient receptor potential cation channels in disease. Physiol Rev 87:165–217

Nishida M, Kurose H (2008) Roles of TRP channels in the development of cardiac hypertrophy. Naunyn-Schmiedebergs Arch Pharmacol 378:395–406

Nowycky MC, Fox AP, Tsien RW (1985) Three types of neuronal calcium channel with different calcium agonist sensitivity. Nature 316:440–3

Ohba T, Watanabe H, Murakami M, Takahashi Y, Iino K, Kuromitsu S, Mori Y, Ono K, Iijima T, Ito H (2007) Upregulation of TRPC1 in the development of cardiac hypertrophy. J Mol Cell Cardiol 42:498–507

Onohara N, Nishida M, Inoue R, Kobayashi H, Sumimoto H, Sato Y, Mori Y, Nagao T, Kurose H (2006) TRPC3 and TRPC6 are essential for angiotensin II-induced cardiac hypertrophy. EMBO J 25:5305–5316

Perez-Reyes E (2003) Molecular physiology of low-voltage activated T-type calcium channel. Nature 391:896–900

Poteser M, Graziani A, Rosker C, Eder P, Derler I, Kahr H, Zhu MX, Romanin C, Groschner K (2006) TRPC3 and TRPC4 associate to form a redox-sensitive cation channel. Evidence for expression of native TRPC3-TRPC4 heteromeric channels in endothelial cells. J Biol Chem 281:13588–95

Poteser M, Schleifer H, Lichtenegger M, Schemthaner M, Stockner T, Kappe CO, Glasnov TN, Romanin C, Groschner K (2011) PKC-dependent coupling of calcium permeation through transient receptor potential canonical 3 (TRPC3) to calcineurin signaling in HL-1 myocytes. Proc Natl Acad Sci USA 108:10556–61

Reuter H (1967) The dependence of slow inward current in Purkinje fibres on the extracellular calcium concentration. J Physiol 192:479–92

Richard S, Leclercq F, Lemaire S, Piot C, Nargeot J (1998) Ca^{2+} currents in compensated hypertrophy and heart failure. Cardiovasc Res 37:300–11

Sandmann S, Spitznagel H, Chung O, Xia QG, Illner S, Jänichen G, Rossius B, Daemen MJ, Unger T (1998) Effects of the calcium channel antagonist mibefradil on haemodynamic and morphological parameters in myocardial infarction-induced cardiac failure in rats. Cardiovasc Res 39:339–350

Sandmann S, Min JY, Meissner A, Unger T (1999) Effects of the calcium channel antagonist mibefradil on haemodynamic parameters and myocardial Ca(2+) handling in infarct-induced heart failure in rats. Cardiovasc Res 44:67–80

Sandmann S, Claas R, Cleutjens JP, Daemen MJ, Unger T (2001) Calcium channel blockade limits cardiac remodeling and improves cardiac function in myocardial infarction-induced heart failure in rats. J Cardiovasc Pharmacol 37:64–77

Satoh S, Tanaka H, Ueda Y, Oyama J, Sugano M, Sugimoto H, Mori Y, Makino N (2007) Transient receptor potential (TRP) protein 7 acts as a G protein-activated Ca^{2+} channel mediating angiotensin II-induced myocardial apoptosis. Mol Cell Biochem 294:205–215

Scheuer J (1999) Catecholamines in cardiac hypertrophy. Am J Cardiol 83:70H–74H

Seth M, Zhang Z-S, Mao L, Graham V, Burch J, Stiber J, Tsiokas L, Winn M, Abramowitz J, Rockman HA, Birnbaumer L, Rosenberg P (2009) TRPC1 channels are critical for hypertrophic signaling in the heart. Circ Res 105:1023–30

Sipido KR, Carmeliet E, Van de Werf F (1998) T-type Ca^{2+} current as a trigger for Ca^{2+} release from the sarcoplasmic reticulum in guinea-pig ventricular myocytes. J Physiol 508(2):439–51

Takebayshi S, Li Y, Kaku T, Inagaki S, Hashimoto Y, Kimura K, Miyamoto S, Hadama T, Ono K (2006) Remodeling excitation-contraction coupling of hypertrophied ventricular myocytes is dependent on T-type calcium channels expression. Biochem Biophys Res Commun 345:766–73

Talavera K, Nilius B (2006) Biophysics and structure-function of T-type Ca^{2+} channels. Cell Calcium 40:97–114

Talavera K, Staes M, Janssen A, Klugbauer N, Droogmans G, Hofmann F, Nilius B (2001) Aspartate residues of the Glu-Glu-Asp-Asp (EEDD) pore locus control selectivity and permeation of the T-type Ca^{2+} channel α1G. J Biol Chem 276:45628–35

Venkatachalam K, Zheng F, Gill DL (2003) Regulation of canonical transient receptor potential (TRPC) channel function by diacylglycerol and protein kinase C. J Biol Chem 278:29031–40

Villame J, Massicotte J, Jasmin G, Dumont L (2001) Effects of mibefradil, a T- and L-type calcium channel blocker, on cardiac remodeling in the UM-X7.1 cardiomyopathic hamster. Cardiovasc Drugs Ther 15:41–48

Wilkins BJ, DeWindt LJ, Bueno OF, Braz JC, Glascock BJ, Kimball TF, Mokentin JD (2002) Targeted disruption of NFATc3, but not NFATc4, reveals an intrinsic defect in calcineurin-mediated cardiac hypertrophic growth. Mol Cell Biol 22:7603–13

Wu X, Zhang T, Bossuyt J, Li X, McKinsey TA, Dedman JR, Olson EN, Chen J, Brown JH, Bers DM (2006) Local InsP3-dependent perinuclear Ca^{2+} signaling in cardiac myocyte excitation-transcription coupling. J Clin Invest 116:675–82

Wu X, Eder P, Chang B, Molkentin JD (2010) TRPC channels are necessary mediators of pathologic cardiac hypertrophy. Proc Natl Acad Sci USA 107:7000–5

Xu M, Welling A, Paparisto S, Hofmann F, Klugbauer N (2003) Enhanced expression of L-type $Ca_v1.3$ calcium channels in murine embryonic hearts from $Ca_v1.2$-deficient mice. J Biol Chem 278:40837–41

Zhou Z, January CT (1998) Both T- and L-type Ca^{2+} channels can contribute to excitation-contraction coupling in cardiac Purkinje cells. Biophys J 74:1830–9

Role of T-Type Ca^{2+} Channels in the Development of Arrhythmias and Ischemia–Reperfusion Injury

6

Stephen W. Schaffer

Abstract

The T-type Ca^{2+} channels are both abundant in pacemaker cells and activated at low thresholds, properties important for involvement in diastolic depolarization. Although T-type Ca^{2+} current is tiny, bradycardia is induced by blocking the current with inhibitors or by genetic manipulation, supporting the view that the T-type Ca^{2+} channels play a role in the regulation of the pacemaker. Besides regulating pacemaker current, the T-type Ca^{2+} channels have been implicated in electrical remodeling and ischemia–reperfusion injury, events involving Ca^{2+} overload.

6.1 Ionic Currents Involved in Diastolic Depolarization

Pacemaker cells located in the sinoatrial (SA) node enable the heart to beat spontaneously at a fixed rate, which can vary from 295 in the mouse to 190 in the rabbit (Cho et al. 2003). Sinoatrial tissue contains primary and secondary pacemaker cells, with the primary pacemaker cells exhibiting significantly higher rates of maximal upstroke velocity (Verheick et al. 2001). The maximal upstroke velocity of the mouse cell is also higher than that of the rat and rabbit, perhaps because of differences in the content of primary pacemaker cells. These pacemaker cells generate periodic action potentials that propagate through the established myocardial conductance system to trigger contractions upon reaching ventricular cardiomyocytes. Several ionic currents contribute to the slow diastolic depolarization of the pacemaker potential, although the importance of each current still remains unclear. Historically, three ionic mechanisms for the underlying basis of

S.W. Schaffer (✉)
Department of Pharmacology, University of South Alabama College of Medicine, Mobile, AL 36688, USA
e-mail: sschaffe@southalabama.edu

S.W. Schaffer, M. Li (eds.), *T-type Calcium Channels in Basic and Clinical Science*,
DOI 10.1007/978-3-7091-1413-1_6

pacemaking have received the most attention. (1) Brown (1982) introduced the "I_K–decay hypothesis," which maintains that the rate of diastolic depolarization is primarily determined by the decay in I_K and an inward background current (I_b). However, the I_K–decay hypothesis was subsequently challenged by DiFrancesco (1993), who found that I_b is an outward directed current rather than the inward directed current expected of transport processes involved in diastolic depolarization. Interestingly, more recently an inward directed sustained current (I_{st}) has been identified in spontaneously beating SA node cells that has been implicated in automaticity (Mitsuiye et al. 2000). (2) DiFrancesco et al. (1991) proposed that a hyperpolarization-activated inward current, termed I_f or funny current, plays a primary role in diastolic depolarization. The funny current is carried by both Na^+ and K^+ ions and is usually activated at threshold potentials of around −40 to −50 mV (Baruscotti et al. 2005; DiFrancesco 2006). The activation curve of I_f is modulated by the autonomic system, with β-adrenergic agonists shifting the activation curve to more positive voltages and acetylcholine having the opposite effect. The primary component responsible for the I_f current of the SA node consists of one or more of the HCN (hyperpolarization-activated cyclic nucleotide- gated) channel family members. HCN1 has the fastest activation kinetics and appears to be the dominant isoform of rabbit SA node (Moroni et al. 2001) while HCN4 has the slowest activation kinetic and serves an important function in diastolic depolarization, not only because it is regulated by cAMP. Interestingly, a large number of the SA nodes of HCN4 knockout mice are quiescent (Baruscotti et al. 2010). HCN2 is present at low to moderate levels in the SA node, raising questions regarding its importance (Baruscotti et al. 2010). (3) The third hypothesis, which is based on the work of Hagiwara et al. (1988), postulates a key role for a slow inward Ca^{2+} current (I_{si}) in SA nodal automaticity. This current consists of both $I_{Ca,L}$ and $I_{Ca,T}$ currents, which represent the current derived from the L-type Ca^{2+} channel and the T-type Ca^{2+} channel, respectively (Mangoni et al. 2006a). Based on the effect of nifedipine, which blocks $I_{Ca,L}$, Verheijck et al. (1999) concluded that flux of Ca^{2+} through the L-type Ca^{2+} channel contributes to diastolic depolarization. There is also evidence supporting a role for $I_{Ca,T}$ in diastolic depolarization, which is a major focus of the present review article.

6.2 Role of L-Type Ca^{2+} Channel in Action Potential Upstroke and Diastolic Depolarization

The dominant inward current of the action potential upstroke of the SA node cells is $I_{Ca,L}$ (Dokos et al. 1996). Inhibition of the L-type Ca^{2+} channel to block $I_{Ca,L}$ also prevents spontaneous action potentials. A secondary increase in $I_{Ca,L}$ is responsible for phase 2 of the action potential, which is terminated by repolarization (phase 3), which is mediated by delayed rectifier K^+ current.

The most abundant Ca^{2+} channel subunit in the heart is $Ca_v1.2$, whose mRNA content is 10–100 times that for $Ca_v2.3$, $Ca_v3.1$, and $Ca_v3.2$ (Larsen et al. 2002). Because of its abundance in the ventricle, $Ca_v1.2$ appears to play a dominant role in providing Ca^{2+} for cardiac contraction, the likely reason its deletion results in death

(Seisenberger et al. 2000; Cho et al. 2003). In mouse SA node, $Ca_v1.2$ is also the major component of the L-type Ca^{2+} channel although its role in pacemaking remains to be established. Although expression of the $Ca_v1.3$ subunit is weak in the SA node (Bohn et al. 2000), deletion of $Ca_v1.3$ gives rise to sinus bradycardia with prolonged PR interval but no fatal phenotype (Platzer et al. 2000; Zhang et al. 2002). Moreover, $Ca_v1.3$ null mice are associated with sinoatrial arrhythmias as well as impaired atrioventricular conduction that includes atrioventricular block (Mangoni et al. 2006a). These findings support the view that, while $Ca_v1.3$ is widely known for its central role in excitation–contraction coupling, it also plays a central role in determining basal heart rate. Interestingly, activation of $I_{Ca,L}$ in SA nodal cells of $Ca_v1.3^{-/-}$ mice is shifted to more depolarizing potentials, suggesting that $Ca_v1.3$ Ca^{2+} channels might contribute to the later phase of diastolic depolarization while $Ca_v1.2$ does not contribute to the depolarization phase (Zhang et al. 2002). These findings are consistent with the observation that inhibition of the L-type Ca^{2+} channel causes bradycardia (Mangoni et al. 2006a). A characteristic feature of the SA node pacemaker is its dependence on Ca^{2+} calmodulin-dependent protein kinase II (CaMKII) activity, an effect attributed to the activation of L-type Ca^{2+} channels (Vinogradova et al. 2000). Moreover, the negative chronotropic effect of vagal stimulation appears to involve hyperpolarization of the maximum diastolic potential (a I_K effect), inhibition of I_f, and muscarinic control of $I_{Ca,L}$ (Zara et al. 1996). However, $Ca_v1.3$ and $Ca_v3.1$ mRNA expressions are elevated in $Ca_v2.3^{-/-}$ mice, indicating that complex interactions exist between the different subunits of L-type and T-type Ca^{2+} channels.

6.3 Role of T-Type Ca^{2+} Channel in Diastolic Depolarization

The threshold of the T-type Ca^{2+} channel of SA nodal cells is -60 mV, and the maximal activation voltage varies from -40 mV to -10 mV (Cho et al. 2003; Ono and Iijima 2005). Thus, not only are T-type Ca^{2+} channels abundant in the SA node, but they are activated at low thresholds (-50 mV) in contrast to the higher threshold (-30 mV) of the L-type Ca^{2+} channel, making them supposedly ideal for diastolic depolarization, which commonly ranges from -65 to -40 mV (Mangoni et al. 2006a). However, the T-type Ca^{2+} channel begins inactivation at -90 mV and is fully inactivated by -40 mV (Ono and Iijima 2005). Hence, $I_{Ca,T}$ has been called the "tiny and transient" current, which prevents it from carrying much inward current to affect pacemaker depolarization.

A key question that remains unanswered is the physiological function of the T-type Ca^{2+} channel. After all, there are more T-type Ca^{2+} channels in the SA node than in any other cardiac cell type. In their pioneer study, Hagiwara et al. (1988) found that treatment of beating SA nodal cells with either Ni^{2+}, a specific inhibitor of the T-type Ca^{2+} channel, or an agent that promotes the release of Ca^{2+} from the sarcoplasmic reticulum during the systolic phase of the Ca^{2+} cycle induces bradycardia, an effect attributed to reduced rates of pacemaker depolarization. Ono and Iijima (2005) have suggested that Ni^{2+} may suppress Ca^{2+} sparks, which ordinarily contribute to pacemaker potential by activating an inward Na^+/Ca^{2+} exchange current. This view was further supported by the observations that other T-type Ca^{2+} channel antagonists, mibefradil and R(-)efonidipine, also suppress the later phase of pacemaker depolarization, with suppression greatest in animals with the highest density of T-type Ca^{2+} channels (Masumiya et al. 1998; Tanaka et al. 2008).

$Ca_v3.1^{-/-}$ mice lack $I_{Ca,T}$ in SA node cells, a change that results in slower rates of pacemaker activity caused by a reduction in the diastolic depolarization slope (Mangoni et al. 2006b). The prolongation of SA node recovery also appears related to diminished T-type Ca^{2+} channel activity. However, maximum diastolic potential, action potential duration, and amplitude and threshold were largely unchanged in $Ca_v3.1^{-/-}$ mice.

6.4 T-Type Ca^{2+} Channels in AV Nodal, Purkinje, and Bundle of His Cells

Both T- and L-type Ca^{2+} channels are present in atrioventricular (AV) nodal cells. Mangoni and coworkers (Mangoni et al. 2006b; Marger et al. 2011) found that $Ca_v1.3^{-/-}$ mice exhibit a severe reduction in AV nodal conductance, accompanied by severe disruption of diastolic depolarization, characterized by subthreshold membrane potential oscillations without spontaneous automaticity. Interestingly, pacemaking could not be restored with isoproterenol in $Ca_v1.3^{-/-}$ mice; thus, $Ca_v1.3$ is required for normal AV nodal conduction and is indispensible to diastolic depolarization, as pacemaker activity in $Ca_v1.3^{-/-}$ mice is completely silent under normal physiological conditions (Platzer et al. 2000; Zhang et al. 2002; Marger et al. 2011). A role for the T-type Ca^{2+} channel in AV conductance has also been proposed, as AV conduction slows in $Ca_v3.1^{-/-}$ mice but not in $Ca_v3.2^{-/-}$ mice (Mangoni et al. 2006b). Moreover, maximal rates of diastolic depolarization are not attainable in $Ca_v3.1^{-/-}$ mice, indicating that $Ca_v3.1$ plays a role in setting maximal pacing rates (Marger et al. 2011). In canine Purkinje fibers, the mRNA for $Ca_v1.2$ is ~100-fold more abundant than that for either $Ca_v3.1$ or $Ca_v3.3$, with only $Ca_v3.2$ present in high levels, suggesting that the T-type Ca^{2+} channel only plays a minor role in the action potential of the cell type (Rosati et al. 2007).

6.5 Role of L-Type and T-Type Ca^{2+} Channels in Atrial Remodeling

Atrial fibrillation is a common arrhythmia among the elderly that tends to become persistent over time. Wijffels et al. (1995) found that episodes of atrial fibrillation induce electrical remodeling of the atria that prolongs the period of atrial fibrillation (atrial fibrillation begets atrial fibrillation). A prominent feature of electrical remodeling induced by rapid stimulation is shortening of action potential duration secondary to the downregulation of the $\alpha 1C$ subunit of the L-type Ca^{2+} channel (Yang et al. 2005). However, in human atrial fibrillation, the decrease in L-type Ca^{2+}

channel activity is associated with a downregulation of all channel subunits, with the greatest decline detected in the α_2 δ subunit (Grammer et al. 2001). The resulting decline in $I_{Ca,L}$ protects the myocyte against Ca^{2+} overload, which is considered the trigger for development of atrial remodeling. Inhibition of the L-type Ca^{2+} channel was found to cause modest to no protection against atrial remodeling while the combination of an L- type and T-type Ca^{2+} channel antagonist was found in several studies to be more effective in preventing atrial pacing-mediated electrical remodeling than treatment with the L-type type Ca^{2+} channel antagonist (Fareh et al. 2001; Ohashi et al. 2004). T-type Ca^{2+} channel inhibition has also been shown to prevent sudden arrhythmic death in patients with chronic heart failure (Kinoshita et al. 2009). However, a recent study challenged the utility of T-type Ca^{2+} channel blockade in preventing atrial fibrillation-related remodeling, as the more selective T-type Ca^{2+} channel antagonist AZ9112 was incapable of preventing atrial fibrillation-mediated remodeling (Kato et al. 2014).

6.6 Role of L-Type and T-Type Ca^{2+} Channels in Ischemic Injury

Calcium overload is a major cause of ischemia–reperfusion injury. During an ischemic insult, elevations in the intracellular concentration of metabolic intermediates, such as lactate, coupled with enhanced hydrolysis of ATP lead to the accumulation of $[H^+]_i$. The resulting decrease in pH_i serves as the driving force behind cellular extrusion of H^+ in exchange for Na^+ via the Na^+/H^+ exchanger. The hydrolysis of ATP and subsequent decline in ATP content has another adverse effect, namely, diminished Na^+–K^+ ATPase activity, which is associated with an elevation in $[Na^+]_i$. The depolarization of the plasma membrane, coupled with the increase in $[Na^+]_i$, stimulates reverse flux through the Na^+/Ca^{2+} exchanger resulting in an increase in $[Ca^{2+}]_i$. Upon reperfusion, the combination of enhanced Na^+ uptake via the Na^+/H^+ exchanger coupled to Ca^{2+} influx via the Na^+/Ca^{2+} exchanger leads to intracellular Ca^{2+} overload. Although a significant increase in $[Ca^{2+}]_i$ can activate several potentially damaging enzymes, such as the protease, calpain (Sandmann et al. 2002), and the hydrolase, phospholipase A_2, the most injurious action of Ca^{2+} overload is severe mitochondrial disruption. The uptake of Ca^{2+} by the mitochondria can trigger the mitochondrial permeability transition to initiate apoptosis (Murphy and Steenbergen 2008).

One of the major causes of mortality during an ischemia–reperfusion insult is the development of ventricular arrhythmias. These arrhythmias are caused by several characteristic changes in the action potential of the cardiomyocyte. First, inhibition of the Na^+–K^+ ATPase, the increase in outward directed K_{ATP} current, and a reduction in $[K^+]_i$ lead to a partial depolarization of the resting membrane potential (Akar and Akar 2007), which alters excitability rendering the cardiomyocyte more susceptible to ectopic beats. The elevation of sympathetic activity also increases the risk of ectopic beats by accelerating further depolarization of the cardiomyocyte. It also diminishes the number of active Na^+ channels, which lowers V_{max} and decreases conduction velocity of the impulse generated upon stimulation, rendering the weakened impulses more susceptible to conduction block. Second, impulse conduction is also adversely affected by the reduction in conduction through impaired gap junctions (Akar and Akar 2007). Third, the decline in pH_i during the ischemic phase of the insult inhibits several ion channels, including both T-type and L-type Ca^{2+} channels (Delise and Satin 2000). However, a more dominant effect is the stimulation of Ca^{2+} influx via the combined actions of the Na^+/H^+ and the Na^+/Ca^{2+} exchangers, events enhanced by acidosis. Ca^{2+} overload is a major cause of arrhythmias, including action potential alternans, a common arrhythmia of the ischemic heart. It can also facilitate the transition from ventricular tachycardia to ventricular fibrillation (Akar and Akar 2007). Fourth, oxidative stress rises in the ischemic heart, but to a much greater degree in the ischemia-reperfused heart (Murphy and Steenbergen 2008). Among other actions, ROS can increase calcium oscillations (Coetzee and Opie 1987). Fifth, catecholamines promote Ca^{2+} overload in the ischemia-reperfused heart by stimulating Ca^{2+} uptake via the L-type Ca^{2+} channel, an effect likely responsible for the antiarrhythmic actions of the β-adrenergic blockers (Lubbe et al. 1992).

It is widely recognized that the L-type Ca^{2+}channel inhibitor, verapamil, prevents the development of arrhythmias during an ischemia–reperfusion insult (Opie et al. 1998). However, the L-type Ca^{2+}channel antagonists also depress contractile function, suppress AV nodal conduction, and cause severe hypotension (Curtis and Walker 1986; Muller et al. 1998). The most intriguing evidence that the T-type Ca^{2+}channel might play a role in post-myocardial infarction injury is based on the reported effects of mibefradil. Using a pig model of acute myocardial ischemia, Muller et al. (1998) studied the effect of mibefradil therapy on the condition of the ischemia-reperfused heart (administered as a bolus of 1 mg/kg body wt 30 min prior to coronary artery ligation and as an infusion of 2 mg/kg/h during the 20 min period of ischemia and during the 25 min reperfusion phase of the insult). They found that mibefradil suppresses ventricular arrhythmias and ventricular fibrillation threshold without diminishing contractile function. The possibility that T-type Ca^{2+}channel blockade could slow heart rate and thereby limit Ca^{2+}-induced arrhythmias was proposed although the dose of mibefradil used does not reduce heart rate. It was also suggested that mibefradil might improve blood flow to the ischemic zone, as it did to the nonischemic zone. The possibility that mibefradil might selectively inhibit L-type Ca^{2+} channels in depolarized tissue was also mentioned by the authors.

In contrast to the study of Muller et al. (1998), Farkas et al. (1999) found mibefradil only exhibited antiarrhythmic activity at a high dosage (600 nM) in the isolated perfused rat heart subject to a regional ischemic insult. Moreover, mibefradil caused a similar degree of AV block and coronary vasodilation as verapamil, suggesting that the antiarrhythmic effect of mibefradil was mediated by its weak action on the L-type Ca^{2+}channel. Yet in a related isolated heart study, Arh and Budihna (2000) found that mibefradil was more effective than the L-type Ca^{2+}channel antagonist, nitrendipine, in diminishing the number of arrhythmic events and lactate dehydrogenase release, a sign of myocardial damage. The same group subsequently found that mibefradil (0.1 μM) was more effective than verapamil (0.1 μM) in preventing ischemia–reperfusion-mediated ventricular fibrillation and in improving recovery of contractile function in the reperfused guinea pig heart. The differences between the mibefradil-treated groups and the verapamil-treated groups were amplified in guinea pigs suffering from acute rental failure, a condition associated with more sustained bouts of ischemia–reperfusion-mediated ventricular fibrillation and contractile failure than the normal guinea pig heart subjected to an ischemia–reperfusion insult (Kuhar et al. 2004). Thus, in a species (guinea pigs) with an abundant number of T-type Ca^{2+}channels, inhibition of the T-type and L-type Ca^{2+}channels with mibefradil is more effective at reducing ischemia–reperfusion injury than inhibition of only the L-type Ca^{2+}channel with the antagonist verapamil. According to Mocanu et al. (1999), mibefradil preconditions the isolated rat heart, an effect abolished by glibenclamide suggesting that the L-type and T-type Ca^{2+}channel blocker may protect the ischemia-reperfused heart by opening the K_{ATP} channel. In agreement with the study of Mocanu et al. (1999), Pastukh et al. (2005) found that preconditioning of the neonatal cardiomyocyte with high glucose medium is prevented by angiotensin II, an effect blocked by the T-type Ca^{2+}channel antagonist, mibefradil, and a Na^+/H^+ exchange antagonist. Thus, elevations in $[Ca^{2+}]_i$ by any number of sources, including those from the L-type Ca^{2+} channel, not only worsen ischemia–reperfusion injury but also antagonize preconditioning. Mozaffari et al. (2006) also found that mibefradil (0.3 μM) specifically attenuates the extension of infarct size mediated by elevations in afterload pressure from 80 to 160 cm H_2O while improving recovery of contractile function and attenuating the increase in end-diastolic pressure of the perfused rat heart. By comparison, the reduction in infarct size by L-type Ca^{2+}channel blockade (diltiazem, 10 μM) was associated with a decrease in contractile function in both the nonischemic and reperfused heart.

Although most studies have reported a beneficial effect of mibefradil against ischemia–reperfusion injury, its limited selectivity has clouded the role of the T-type Ca^{2+} channel in the improvement of tissue damage. This limitation is overcome in mouse lines that overexpress or downregulate subunits of the T-type Ca^{2+}channel, $Ca_v3.1$ and $Ca_v3.2$. Using this approach, Quang et al. (2011) found no difference in infarct size between either $Ca_v3.1$ and $Ca_v3.1^{-/-}$ or $Ca_v3.2$ and $Ca_v3.2^{-/-}$ measured 4 weeks after occlusion of the left anterior descending coronary artery. However, the method used to measure infarct size in their study did not employ the standard method of determining infarct size, area at risk. Rather, the infarct size data merely indicated that the area at risk was identical in the two experimental groups. Hence, the major conclusion of the study is that $Ca_v3.1$ plays a role in postischemic ventricular remodeling while $Ca_v3.2$ is without any effect on remodeling. Pastukh et al. (2010) also observed that downregulation of the $Ca_v3.1$ T-type Ca^{2+} channel attenuated energy deficiency-mediated Ca^{2+} accumulation and cell death in neonatal cardiomyocytes incubated with medium containing metabolic inhibitors (deoxyglucose and the electron transport chain inhibitor amobarbital).

The authors proposed that the $Ca_v3.1$ T-type Ca^{2+} channel plays a role in ischemia–reperfusion injury.

Conclusions

Traditionally, three mechanisms were thought to contribute to pacemaker current: I_K decay, I_f, and I_{Ca}. Calcium current is derived from both the L-type and T-type Ca^{2+} channels, with the $Ca_v1.3$ and $Ca_v3.1$ subunits playing the most important roles in diastolic depolarization. Several studies have concluded that both the L-type and T-type Ca^{2+} channels play a role in atria fibrillation-induced atrial remodeling, as antagonists blocking both types of Ca^{2+} channels are more effective than those that just block the L-type Ca^{2+} channel. However, a recent study using a more specific T-type Ca^{2+} channel antagonist (AZ9112) challenges this notion, raising the need for further studies.

Calcium overload is an important mechanism of ischemia–reperfusion injury, including the development of ventricular arrhythmias. Several mechanisms contribute to the accumulation of excessive levels of intracellular Ca^{2+} during an ischemia–reperfusion insult, including the Na^+/Ca^{2+} exchanger, the L-type Ca^{2+} channel, and the T-type Ca^{2+} channel. Interest in the use of T-type Ca^{2+} channel antagonists arose because the L-type Ca^{2+} channel antagonists, while effective at diminishing tissue injury and the development of ventricular arrhythmias, also decrease contractile function, in contrast to T-type Ca^{2+} channel antagonists that can benefit the ischemia-reperfused heart without decreasing contractile function. The T-type Ca^{2+} channel antagonist, mibefradil, has the added benefit of protecting the heart by preconditioning it. The major T-type Ca^{2+} channel subunit implicated in ischemia–reperfusion injury is $Ca_v3.1$, although further studies examining the contribution of $Ca_v3.2$ are warranted. Because Ca^{2+} overload is a major cause of myocardial damage during both the ischemic and reperfusion phases of an ischemia–reperfusion insult, clarification of the contribution of the T-type Ca^{2+} channel in promoting ischemic and reperfusion damage has important clinical implications. It is attractive to consider that selective T-type Ca^{2+} channel antagonists might serve an important role in diminishing Ca^{2+}-mediated cardiac injury while preventing the loss of contractile function.

References

Akar JG, Akar FG (2007) Regulation of ion channels and arrhythmias in the ischemic heart. J Electrocardiol 40:S37–S41

Arh M, Budihna MV (2000) Comparison of effects of nitrendipine, lacidipine and mibefradil on postischaemic myocardial damage in isolated rat hearts. Pflugers Arch 440(Suppl 5):R149–R150

Baruscotti M, Bucchi A, DiFrancesco D (2005) Physiology and pharmacology of the cardiac pacemaker ("funny") current. Pharmacol Ther 107:59–79

Baruscotti M, Barbuti A, Bucchi A (2010) The cardiac pacemaker current. J Mol Cell Cardiol 48:55–64

Bohn G, Moosmang S, Conrad H, Ludwig A, Hofmann F, Klugbauer N (2000) Expression of T-type and L-type Ca^{2+} channel mRNA in murine sinoatrial node. FEBS Lett 481(1):73–76
Brown H (1982) Electrophysiology of the sinoatrial node. Physiol Rev 62:505–530
Cho H-S, Takano M, Noma A (2003) The electrophysiological properties of spontaneously beating pacemaker cells isolated from mouse sinoatrial node. J Physiol 550:169–180
Coetzee WA, Opie LH (1987) Effects of components of ischaemia and metabolic inhibition on delayed after depolarizations in guinea pig papillary muscle. Circ Res 61:157–165
Curtis MJ, Walker MJA (1986) The mechanism of action of the optical enantiomers of verapamil against ischaemia-induced arrhythmias in the conscious rat. Br J Pharmacol 89:137–147
Delise BP, Satin J (2000) pH modification of human T-type calcium channel gating. Biophys J 78:1895–1905
DiFrancesco D (1991) The contribution of the 'pacemaker' current (If) to generation of spontaneous activity in rabbit sino-atrial node myocytes. J Physiol (Lond) 434:23–40
DiFrancesco D (1993) Pacemaker mechanisms in cardiac tissue. Annu Rev Physiol 55:455–472
DiFrancesco D (2006) Serious workings of the funny current. Prog Biophys Mol Biol 90:13–25
Dokos S, Celler B, Lovell N (1996) Ion currents underlying sinoatrial node pacemaker activity: a new single cell mathematical model. J Theor Biol 181:245–272
Fareh S, Benardeau A, Nattel S (2001) Differential efficacy of L- and T-type calcium channel blockers in preventing tachycardia-induced atrial remodeling in dogs. Cardiovasc Res 49:762–770
Farkas A, Qureshi A, Curtis MJ (1999) Inadequate ischaemia-selectivity limits the antiarrhythmic efficacy of mibefradil during regional ischaemia and reperfusion in the rat isolated perfused heart. Br J Pharmacol 128:41–50
Grammer JB, Zeng X, Bosch RF, Kuhlkamp V (2001) Atrial L-type Ca2+ channel, β-adrenoceptor, and 5-hydroxytryptamine type 4 receptor mRNAs in human atrial fibrillation. Basic Res Cardiol 96:82–90
Hagiwara N, Irisawa H, Kameyama M (1988) Contribution of two types of calcium currents to the pacemaker potentials of rabbit sinoatrial node cells. J Physiol 550:169–180
Kato T, Iwasaki YK, Duker G, Fjellstrom O, Giordanetto F, Sundqvist M, Wallin A, Wang Q-D, Nattel S (2014) Inefficiency of a highly selective T-type calcium channel blocker in preventing atrial fibrillation related remodeling. J Cardiovasc Electrophysiol 25:531–536
Kinoshita H, Kuwahara K, Takano M, Arai Y, Kuwabara Y, Yasuno S, Nakagawa Y, Nakanishi M, Harada M, Fujiwara M, Murakami M, Ueshima K, Nakao K (2009) T-type Ca^{2+} channel blockade prevents sudden death in mice with heart failure. Circulation 120:743–752
Kuhar CG, Budihna MV, Pleskovic RZ (2004) Mibefradil is more effective than verapamil for restoring post-ischemic function of isolated hearts of guinea pigs with acute renal failure. Eur J Pharmacol 488:137–146
Larsen JK, Mitchell JW, Best PM (2002) Quantitative analysis of the expression and distribution of calcium channel α1 subunit mRNA in the atria and ventricles of the rat heart. J Mol Cell Cardiol 34:519–532
Lubbe WF, Podzuweit T, Opie LH (1992) Potential arrhythmogenic role of cyclic adenosine monophosphate (AMP) and cytosolic calcium overload: Implications for prophylactic effects of beta-blockers in myocardial infarction and proarrhythmic effects of phosphodiesterase inhibitors. J Am Coll Cardiol 19:1622–1633
Mangoni ME, Couette B, Marger L, Bourinet E, Striessnig J, Nargeot J (2006a) Voltage-dependent calcium channels and cardiac pacemaker activity: from ionic current to genes. Prog Biophys Mol Biol 90:38–63
Mangoni ME, Traboulsie A, Leoni A-L, Couette B, Marger L, Quang KL, Kupfer E, Cohen-Solal A, Vilar J, Shin H-S, Escande D, Charpentier F, Nargeot J, Lory P (2006b) Bradycardia and slowing of the atrioventricular conduction in mice lacking $Ca_v3.1/\alpha_{1G}$ T-type calcium channels. Circ Res 98:1422–1430

Marger L, Mesirca P, Alig J, Torrente A, Dubel S, Engeland B, Kanani S, Fontanaud P, Striessnig J, Shin H-S, Isbrandt D, Ehmke H, Nargeot J, Mangoni ME (2011) Function roles of $Ca_v1.3$, $Ca_v3.1$ and HCN channels in automaticity of mouse atrioventricular cells: Insights into the atrioventricular pacemaker mechanism. Channels 5:251–261

Masumiya H, Shijuku T, Tanaka H, Shigenobu K (1998) Inhibition of myocardial L- and T-type Ca^{2+} currents by efonidipine: possible mechanism for its chronotropic effect. Eur J Pharmacol 349:351–357

Mitsuiye T, Shinagawa Y, Noma A (2000) Sustained inward current during pacemaker depolarization in mammalian sinoatrial node cells. Circ Res 87:88–91

Mocanu MM, Gadgil S, Yelolon DM, Baxster GF (1999) Mibefradil, a T-type and L-type calcium channel blocker, limits infarct size through a glibenclamide-sensitive mechanism. Cardiovasc Drugs Ther 13:115–122

Moroni A, Gorza L, Beltrame M, Gravante B, Vaccari T, Bianchi ME, Altomare C, Longhi R, Heurteaux C, Vitadello M, Malgaroli A, DiFrancesco D (2001) Hyperpolarization-activated cyclic nucleotide-gated channel 1 is a molecular determinant of the cardiac pacemaker current I_f. J Biol Chem 276:29233–29241

Mozaffari MS, Patel C, Schaffer SW (2006) Mechanisms underlying after load-induced exacerbation of myocardial infarct size: Role of T-type Ca^{2+} channel. Hypertension 47:912–919

Muller CA, Opie LH, McCarthy J, Hofmann D, Pineda CA, Peisach M (1998) Effects of mibefradil, a novel calcium channel blocking agent with T-type activity, in acute experimental myocardial ischemia: Maintenance of ventricular fibrillation threshold without inotropic compromise. J Am Coll Cardiol 32:268–274

Murphy E, Steenbergen C (2008) Mechanisms underlying acute protection from cardiac ischemia-reperfusion injury. Physiol Rev 88:581–609

Ohashi N, Mitamura H, Tanimoto K, Fukuda Y, Kinebuchi O, Kurita Y, Shiroshita-Takeshita A, Miyoshi S, Hara M, Takatsuki S, Ogawa S (2004) A comparison between calcium channel blocking drugs with different potencies for T- and L-type channels I preventing atrial electrical remodeling. J Cardiovasc Pharmacol 44:386–392

Ono K, Iijima T (2005) Pathophysiological significance of the T-type Ca^{2+} channels: Properties and functional roles of T-type Ca^{2+} channels in cardiac pacemaker. J Pharmacol Sci 99:197–204

Opie LH, Coetzee WA, Dennis SC, Thandroyen FT (1998) A potential role for calcium ions in early ischemic and reperfusion arrhythmias. Ann N Y Acad Sci 522:464–477

Pastukh V, Wu S, Ricci C, Mozafffari M, Schaffer S (2005) Reversal of hyperglycemic preconditioning by angiotensin II: role of calcium transport. Am J Physiol 288:H1965–H1975

Pastukh V, Chen H, Wu S, Jong CJ, Alexeyev M, Schaffer SW (2010) Effect of hypernatremia on injury caused by energy deficiency: role of T-type Ca^{2+} channel. Am J Physiol 299:C289–C297

Platzer J, Engel J, Schrott-Fischer A, Stephan K, Bova S, Chen H, Zheng H, Stiessnig J (2000) Congenital deafness and sinoatrial node dysfunction in mice lacking class D L-type Ca^{2+} channels. Cell 102:89–97

Quang KL, Naud P, Qi XY, Duval F, Shi Y-F, Gillis M-A, Comtois P, Tardif J-C, Li D, Levesque PC, Dobrev D, Charpentier F, Nattel S (2011) Role of T-type calcium channel subunits in post-myocardial infarction remodelling probed with genetically engineered mice. Cardiovasc Res 91:420–428

Rosati B, Dun W, Hirose M, Boyden PA, McKinnon D (2007) Molecular basis of the T-type and L-type Ca^{2+} channels in canine Purkinje fibres. J Physiol 579:465–471

Sandmann S, Spormann J, Prenzel F, Shaw L, Unger T (2002) Calcium channel blockade limits transcriptional, translational and functional up-regulation of the cardiac calpain system after myocardial infarction. Eur J Pharmacol 453:99–109

Seisenberger C, Specht V, Welling A, Platzer J, Pfeifer A, Kuhbandner S, Striessnig J, Klugbauer N, Feil R, Hofmann R (2000) Functional embryonic cardiomyocytes after disruption of the L-type $\alpha1C$ (Cav1.2) calcium channel gene in the mouse. J Biol Chem 275:39193–39199

Tanaka H, Komikado C, Namekata I, Nakamura H, Suzuki M, Tsuneoka Y, Shigenobu K, Takahara A (2008) Species difference in the contribution of T-type calcium current to cardiac pacemaking as revealed by R(-)-efonidipine. J Pharmacol Sci 107:99–102

Verheick EE, Van Kempen MJA, Veereschild M, Lurvink J, Jongsma HJ, Bouman LN (2001) Electrophysiological features of the mouse sinoatrial node in relation to connexin distribution. Cardiovasc Res 52:40–50

Verheijck EE, Van Ginneken ACG, Wilders R, Bouman LN (1999) Contribution of L-type Ca^{2+} current to electrical activity in sinoatrial nodal myocytes of rabbits. Am J Physiol 276:H1064–H1077

Vinogradova TM, Zhou Y-Y, Bogadanov KY, Yang D, Kuschel M, Cheng H, Xiao R-P (2000) Sinoatrial node pacemaker activity requires Ca^{2+}-calmodulin-dependent protein kinase II activation. Circ Res 87:760–767

Wijffels MC, Kirchhof C, Dorland R, Allessie MA (1995) Atrial fibrillation begets atrial fibrillation: a study in awake chronically instrumented goats. Circulation 92:1954–1968

Yang Z, Shen W, Rottman JN, Wikswo JP, Murray KT (2005) Rapid stimulation causes electrical remodeling in cultured atrial myocytes. J Mol Cell Cardiol 38:299–308

Zara A, Robinson RB, DiFrancesco D (1996) Basal responses to the L-type Ca^{2+} hyperpolarization-activated currents to autonomic agonists in the rabbit sino-atrial node. J Physiol 491:347–355

Zhang Z, Xu Y, Song H, Rodriguez J, Tuteja D, Namkung Y, Shin H-S, Chiamvimonvat N (2002) Functional roles of $Ca_v1.3$ (α_{1D}) calcium channel in sinoatrial nodes: insight gained using gene-targeted null mutant mice. Circ Res 90:981–987

Link Between Absence Seizures and T-Type Calcium Channels

7

Yucai Chen and W. Davis Parker

Abstract

An animal model of human absence epilepsy containing a G to C mutation of the $Ca_v3.2$ T-type Ca^{2+} channel gene (*Cacna1h*) ties together the gene mutation, increased T-type Ca^{2+} channel activity and the epileptic phenotype. Mice lacking a related gene (*Cacna1a*) also show enhanced T-type Ca^{2+} current and increased susceptibility to absence seizures. On the other hand, mutations that decrease T-type Ca^{2+} channel activity in thalamocortical relay neurons display no spike–wave discharges associated with absence seizures. These animal models are supported by genetic studies showing defects in T-type Ca^{2+} channel function in humans suffering from epilepsy. Thus, in both human and animal studies, T-type Ca^{2+} channel antagonists show promise in the treatment of absence seizures.

7.1 Introduction

Epilepsy is a disorder of the brain characterized by an enduring predisposition to generate epileptic seizure and by the neurobiological, cognitive, psychological, and social consequences of this condition (Fisher et al. 2005). An epileptic seizure is defined as a transient symptom of abnormal excessive or synchronous neuronal activity in the brain (Fisher et al. 2005). Absence seizures are ascribed to a brief, sudden lapse of consciousness and are considered one of the more common epileptic seizures. Impairment of consciousness and EEG generalized spike–slow wave discharges (SWDs) are two essential characteristics of absence seizure (Panayiotopoulos et al. 1989).

The neurons in the reticular thalamic nucleus (RT), thalamic relay neurons (thalamocortical (TC) neurons), and neocortical pyramidal cells comprise a circuit

Y. Chen (✉) • W.D. Parker
Department of Pediatrics, University of Illinois at Chicago, Peoria, IL 61605, USA
e-mail: yucaic@uicomp.uic.edu

S.W. Schaffer, M. Li (eds.), *T-type Calcium Channels in Basic and Clinical Science*,
DOI 10.1007/978-3-7091-1413-1_7

that participates in the formation of sleep. An alteration in this circuitry is believed to be involved in the mechanism of spindles and spike–wave discharges (SWD) and absence seizures (Steriade 1974; Steriade and Contreras 1995; Futatsugi and Riviello 1998; Timofeev et al. 1998; Fuentealba and Steriade 2005; Pinault and O'Brien 2005). Studies have shown that thalamocortical circuits govern the rhythm of cortical excitation by the thalamus and underlie normal physiologic patterns such as those that occur during sleep. The thalamic relay neurons can activate the cortical pyramidal neurons either in a tonic mode, which occurs during wakefulness and rapid-eye-movement (REM) sleep, or in a burst mode, which occurs during non-REM sleep (Steriade et al. 1993). The burst mode is created by T-type calcium channels, which allow for low-threshold depolarization that generates bursts of action potential when voltage-gated sodium channels are generated (Perez-Reyes et al. 1998). The character of tonic or burst firing in the thalamocortical circuit is primarily controlled by input from the thalamic reticular neurons, which hyperpolarize the relay neurons, allowing them to fire in bursts (Chang and Lowenstein 2003). In the normal awake state, the thalamic relay neurons fire in the tonic mode, and thalamocortical projections transfer sensory information to the cortex in a nonrhythmic manner. However, in absence seizures, the abnormal circuit causes rhythmic activation of the cortex during wakefulness and results in SWD and clinical manifestations of impairment of consciousness (Kostopoulos 2001).

Three different T-type calcium channels express T-type currents: alpha 1G (*Cacna1g*, Cav3.1), alpha 1H (*Cacna1h*, Cav3.2), and alpha 1I (*Cacna1i*, Cav3.3) (Cribbs et al. 1998; Perez-Reyes et al. 1998, 1999; Lee et al. 1999; Perez-Reyes 2003). Considerable data suggest that abnormal T-type current may be the primary culprit in the formation of absence seizures: (1) mutations of the human and mouse Cacna1a calcium channel gene, mutations of the Cacna1a's ancillary genes, and knockout of the Cacna1a initiate a complex absence-associated epileptic condition that is associated with an ataxia phenotype. In these mutated mice, elevated neuronal low-voltage-activated T-type calcium currents influence thalamocortical network activity and contribute to the generation of cortical spike–wave discharges (SWDs) associated with absence seizures (Zhang et al. 2002, 2004; Ernst et al. 2009). (2) Genetic absence epileptic rats from Strasbourg (GAERS), which present with recurrent generalized nonconvulsive seizures characterized by bilateral and synchronous spike–wave discharges (SWDs) and accompanied with behavioral arrest and staring (Marescaux and Vergnes 1995), exhibit elevated thalamic T-type current and increased cacna1g and Cacna1h mRNA expression (Tsakiridou et al. 1995; Talley et al. 2000). More recently, data have shown that a Cav3.2 T-type calcium channel point mutation mediates splice-variant-specific effects on function, and the mutation correlates directly with the epileptic phenotype in the GAERS animal model (Powell et al. 2009). (3) Overexpression of Cacna1g in mice results in spike–wave discharges and pure absence seizures (Ernst et al. 2009), while mice lacking Cacna1g T-type calcium channels exhibit no burst firings in thalamic relay neurons and are resistant to absence seizures (Kim et al. 2001; Song et al. 2004). (4) Gain-of-function polymorphisms in the human *Cacna1h* gene prevail in patients with absence seizures (Chen et al. 2003a;

Khosravani et al. 2004, 2005; Lu et al. 2005; Vitko et al. 2005; Peloquin et al. 2006; Heron et al. 2007). Moreover, functional polymorphisms in the *CACNA1G* gene have been also detected in a related juvenile absence syndrome (Singh et al. 2007). (5) Full antagonism of a T-type calcium channel can inhibit absence seizures and reduce the duration and cycle frequency of SWD (Barton et al. 2005; Tringham et al. 2012).

7.2 Low-Threshold Calcium Current from T-Type Calcium Channels Gene Is Involved in the Formation of Absence Seizures

The major characterizations of T-type channels include low-voltage thresholds for activation and inactivation, fast inactivation, and small single channel conductance in isotonic Ba^{2+} (Perez-Reyes et al. 1999). In neurons, calcium influx through T-type calcium channels triggers low-threshold spikes, which in turn activates a burst of action potentials mediated by sodium channels (burst firing). Burst firing is thought to play an important role in the synchronized activity of the thalamus observed in absence epilepsy (Perez-Reyes 2003).

7.2.1 Calcium-Dependent Burst Firing, Driven by T-Type Calcium Currents, Is Necessary for the Genesis of Spike–Wave Discharge of Absence Seizures

Cortical neurons directly innervate reticular and thalamocortical neurons, whereas the reticular neurons (RT) provide GABAergic projections onto each other and onto thalamocortical neurons, which in turn synapse onto neocortical neurons. The thalamus receives most sensory input; the primary source of excitatory synapses projecting to the thalamus is from the cortex (Liu et al. 1995; Erisir et al. 1997a, b; Khosravani and Zamponi 2006). Both neocortical and thalamocortical cells have excitatory projections back to reticular neurons. The activity of the circuit is further regulated via thalamic local circuits and neocortical interneurons (Steriade et al. 1993). The circuit containing the reticular thalamic nucleus (RT), thalamic relay neurons, and neocortical pyramidal cells has been implicated in the formation of sleep spindles and spike–wave discharges and is involved in the mechanism of absence epilepsy (Futatsugi and Riviello 1998). Thalamic neurons fire in two different modes, burst or tonic (Steriade et al. 1993; Ramcharan et al. 2000; Sherman 2001). The state of the neurons determines the mode of firing: inhibition-induced hyperpolarization leads to burst firing, while excitation-mediated depolarization leads to tonic firing (Steriade et al. 1993). T-type calcium channels underlie burst firings (Cain and Snutch 2013). The entry of calcium ions through T-type calcium channels leads to the depolarization of the membrane, allowing T-type current to generate low-threshold spikes (LTS) that trigger bursts of sodium-dependent action potentials (Perez-Reyes 2003; Cain and Snutch 2010).

However, T-type calcium channels remain inactive when the membrane potential is in its resting state. The channels must first recover from inactivation by membrane hyperpolarization below the resting potential. They then can be activated by small depolarizations driven by hyperpolarization-activated current (IH currents). When activated, T-type channels produce low-threshold calcium currents which induce rebound excitation above threshold that in turn triggers the generation of a burst of action potentials. Therefore, inhibitory inputs are essential for burst firings by neurons, which play a central role in the pathogenesis of absence epilepsy (McCormick and Bal 1997; Perez-Reyes 2003).

RT neurons are regarded as pacemakers of spindles and play a central role in spindle oscillations involved in early sleep (Steriade et al. 1985, 1986, 1987; Fuentealba and Steriade 2005). They are also thought to inhibit thalamocortical neurons during cortically generated spike–wave (absence) seizures, which may explain the obliteration of input from external stimuli and unconsciousness during epileptic fits (Fuentealba and Steriade 2005). More recent data have shown that an alteration in the firing properties of thalamocortical relay (TC) nuclei, caused by a disruption of a Cacna1g single gene in mice, is sufficient to induce absence seizures (Kim et al. 2001; Song et al. 2004; Ernst et al. 2009).

7.2.2 Absence Seizure Animal Model GAERS Has Alternative Alpha1 H Transcripts and Selectively Increases T-Current in RT Neurons

In genetic absence epileptic rats from Strasbourg (GAERS), 100 % of the rats present with recurrent generalized nonconvulsive seizures characterized by bilateral and synchronous spike–wave discharges (SWD) accompanied by behavioral arrest, staring, and sometimes twitching of the vibrissae. Spontaneous SWD (7–11 cps) start and end abruptly on a normal background EEG at a mean frequency of 1.5 per minute. In GAERS, drugs effective against absence seizures in humans suppress the SWD in a dose-dependent manner, whereas drugs specific for convulsive or focal seizures are ineffective (Marescaux and Vergnes 1995). The many similarities between GARES and human absence seizures support using this type of genetic rodent model, as it likely plays an important role in understanding the causes of human absence seizures. Early research showed that T-type current is selectively increased in the RT neurons of GAERS (Tsakiridou et al. 1995). A further study using a quantitative in situ hybridization technique demonstrated a significant, albeit small, elevation in T-type calcium channel mRNA (alpha 1G and alpha 1H) in the thalamus of GAERS (Talley et al. 2000). Moreover, a mutation of alpha 1H gene in GAERS was discovered, which correlates with the number and frequency of seizures in progeny of an F1 intercross. This mutation, R1584P, is located in a portion of the III–IV linker region in Cav3.2 at Exon24. There are two major thalamic Cacna1h splice variants in GAERS, either with or without Exon25 introduced into the splice. The variants act "epistatically" and require the presence of Exon25 to produce significantly fast recovery from channel inactivation and great charge transference during high-frequency bursts. Of particular interest is that the ratio of Cav3.2 (+25 Exon) mRNA to Cav3.2 (−25 Exon) mRNA is greater in the thalamus of GAERS animals compared with non-epileptic controls, which suggests that the relative proportion of Cav3.2 (+25) to Cav3.2 (−25) is subject to transcriptional regulation (Powell et al. 2009).

7.2.3 The Absence Seizures Associated with Mutations of Cacna1a, and Cacna1a's Ancillary Calcium Channel Subunits in Mice Are the Results of Indirect Potentiating T-Type Calcium Current

The Cacna1a gene encodes the transmembrane pore-forming subunit of the P-/Q-type or CaV2.1 voltage-gated calcium channel, which are the principal channels supporting neurotransmitter release in the mammalian central nervous system (Westenbroek et al. 1995). Cacna1a genes are profoundly expressed in the cell bodies and dendrites of cerebellar Purkinje and granule cells (Westenbroek et al. 1995). Dominant mutations in Cacna1a are associated with episodic ataxia type 2, familial hemiplegic migraine type 1, and spinocerebellar ataxia type 6, and 7 % have absence epilepsy (Rajakulendran et al. 2010, 2012). Several spontaneously occurring homozygous mouse mutants of Cacna1a are good research models of human absence epilepsy, having provided information on tottering, leaner, rocker, rolling Nagoya, lethargic, ducky, and stargazer (Fletcher et al. 1996; Burgess and Noebels 1999a, b; Fletcher and Frankel 1999). These strains exhibit episodes of motor arrest with spike–wave EEG similar to that seen in human absence epilepsy but also show cerebellar degeneration, ataxia, and dystonia. It has been observed that a 45 % increase in peak current densities of T-type calcium channel currents is evoked at −50 mV from −110 mV in tottering (Cav2.1/alpha 1A subunit), lethargic (β4 subunit), and stargazer (γ4 subunit) mice compared with wild type. The half-maximal voltages for steady-state inactivation of T-type calcium channel currents were shifted in a depolarized direction by 7.5–13.5 mV in these three mutants (Zhang et al. 2002); these data demonstrate that a mutation in Cacna1a or in its regulatory subunit genes increases intrinsic membrane excitability in thalamic neurons by potentiating T-type calcium channel currents. Another example of Cacan1a mutations that correlate with absence seizure comes from the Cacna1a knockout mice. Mice with a null mutation of Cacna1a (alpha1A−/−) are susceptible to absence seizures characterized by typical spike–wave discharges (SWDs) and behavioral arrests. Isolated thalamocortical relay (TC) neurons from these knockout mice show increased T-type calcium currents in vitro (Song et al. 2004).

7.2.4 Mice with Genetically Modified T-Type Calcium Channels Show Aberrant t-Current Implicated in the Genesis of Absence Seizures

7.2.4.1 Cav3.1 Genetically Modified Mice

Mice with a null mutation of the alpha 1G subunit of the T-type calcium channel lack the ability to generate burst mode action potentials but show the normal pattern of tonic mode firing. The thalamus in these alpha 1G-deficient mice is specifically resistant to the generation of spike–wave discharges in response to GABA (B) receptor activation (Kim et al. 2001). Therefore, alpha 1G T-type calcium channels are thought to play a critical role in the genesis of absence seizures in the thalamocortical pathway by modulation of the intrinsic firing pattern (Kim et al. 2001). This idea was further supported by Cacna1a (−) and Cacna1g (−) double mutation mice, which demonstrate that generation of SWDs in mutant Cacna1a is suppressed by the deletion of the Cacna1g gene. Cross-breeding alpha 1A−/− mice with mice harboring a null mutation of alpha 1G show a complete loss of T-type calcium current in TC neurons and display no SWDs. Similar results were obtained using double-mutant mice harboring the alpha 1G mutation plus another mutation such as lethargic (beta4(lh/lh)), tottering (alpha1A(tg/tg)), or stargazer (gamma2(stg/stg) (Song et al. 2004). More importantly, two BAC transgenic murine lines which overexpress the Cacna1g gene induce pure absence epilepsy through genetic enhancement of the thalamocortical network (Ernst et al. 2009).

7.2.4.2 Knockout of Cav3.2

In Cav3.2 knockout (KO) mice, the burst of RT neurons has a lower spike frequency and less prominent acceleration-deceleration change. In contrast, ventroposterior neurons (VP) of Cav3.2 KO mice showed a higher ratio of bursts and a higher discharge rate within a burst than those of the wild-type (WT) control. In addition, the long-lasting tonic episodes in RT neurons of the Cav3.2 KO have less stereotypic regularity than episodes in their WT counterparts (Liao et al. 2011). Another example of cacna1h gene's implication in epileptogenesis originated from the observation of the pilocarpine model of epilepsy. According to one report, there is a transient and selective upregulation of Cav3.2 subunits at the mRNA and protein levels after pilocarpine-induced status epilepticus. These functional changes are absent in mice lacking Cav3.2 subunits. Essentially, the development of neuropathological hallmarks of chronic epilepsy, such as subfield-specific neuron loss in hippocampal formation and mossy fiber sprouting, is almost completely absent in Cav3.2 knockout mice (Becker et al. 2008).

7.2.4.3 Knockout of Cav3.3

There are two T-type calcium channel genes expressed in the nucleus reticularis thalami (RT), Cav3.2 and Cav3.3, with the Cav3.3 protein being more abundantly expressed. In the transgenic CaV3.3−/− murine line, the absence of Cav3.3 channels in RT cells prevented oscillatory bursting in the low-frequency (4–

10 Hz) range but spared tonic discharge. In contrast, adjacent TC neurons expressing Cav3.1 channels retain low-threshold bursts (Astori et al. 2011).

7.2.5 Antiseizure Drugs and T-Type Calcium Channel

T-type calcium channels play an important role in the generation and maintenance of SWDs in absence seizures. Decreasing the functions of alpha 1G, alpha 1H, and alpha 1I through the use of antiepileptic drugs may play a role in reducing absence seizures by decreasing excitability in thalamic circuits and the ability to recruit network oscillatory activity in thalamic and thalamocortical circuits. Ethosuximide (ETX), a known antiepileptic drug effective in the treatment of generalized absence seizures, was shown to block T-type calcium channels in thalamic relay neurons (Leresche et al. 1998; Barton et al. 2005; Broicher et al. 2007). A new study identified two T-type calcium channel blockers, Z941 and Z944, with attenuated burst firing of thalamic reticular nucleus neurons in GAERS. Z941 and Z944 suppress absence seizures by 85–90 % and reduce both the duration and the cycle frequency of SWDs in GAERS. It has been suggested that Z941 and Z944 likely target the predominant neural circuitry involved in SWDs by inhibiting the ictogenic properties of the cortical neurons, as well as by disrupting the resonant circuitry of the thalamocortical and RT neurons (Tringham et al. 2012). The ability of the T-type calcium channel antagonists to inhibit absence seizures and reduce the duration and cycle frequency of spike–wave discharges also suggests that T-type current generated by T-type calcium channels is a key component in the formation of absence seizures.

7.3 Genetics of Absence Seizure

Twin studies have provided convincing evidence implicating genetic influences in the development of absence seizures. The concordance for the presence of childhood absence epilepsy is 70–85 % and 33 % in monozygotic twins and first-degree relatives, respectively (Crunelli and Leresche 2002). Therefore, idiopathic epilepsy is regarded as primarily genetic but may be polygenic, with different variant alleles working together to contribute to the symptoms of epilepsy (Mulley et al. 2005). Over the past decade, a number of genes have been associated with rare monogenic and idiopathic epilepsies that have relatively simple inheritance. These studies revealed the importance of ion channel genes in epilepsy, but these mutations occur only in rare monogenic epilepsy syndrome (Lu and Wang 2009) and therefore may not be applicable to common seizures. The complex genetics of idiopathic epilepsy, which involves multiple alleles working together, means that the traditional method of seeking pathological genes through genetic linkage may not be a good strategy in this type of disease. Direct sequencing of candidate genes for rare functional variants based, focusing on selected mechanisms to identify the susceptibility genes of absence seizure, is a better option (Lowenstein and Messing 2007).

Patients with childhood absence epilepsy (CAE) have the following specific characteristics: (1) onset in childhood, which may remit in the adult; (2) brief absence seizures (~4–20 s) that occur frequently, sometimes hundreds per day, characterized by 3 Hz spike wave on their EEG; (3) the symptoms of CAE can be activated by hyperventilation or light stimulation; (4) it is genetically predisposed, with a 16–45 % positive family history; (5) T-type calcium channel blockers, which attenuate thalamic burst firing and suppress absence seizures, are very effective drugs in treating patients with CAE; and (6) no abnormal radiological finding, and patients are otherwise neurologically normal (Porter 1993; Crunelli and Leresche 2002). These unique characteristics of CAE suggest that there are some genes that correlate with the maintenance of thalamocortical synchronization and both cognitive and brain development may be improperly expressed. The gene's expression may be increased in the early developmental stage of the brain but decrease or become silent in the adult. Environmental factors may also affect gene expression. Research information indicates that the circuit neurons of the reticular thalamic nucleus (RT), thalamic relay cells, and neocortical pyramidal cells comprise a circuit implicated in the formation of sleep spindles and spike–wave discharges and the mechanisms of absence epilepsy. Therefore, factors that interfere with this circuit may lead to absence seizures. The mice of alpha (1G)-deficient thalamus are specifically resistant to the generation of spike–wave discharges in response to GABA (B) receptor activation (Kim et al. 2001). On the other hand, transgenic murine lines overexpressing the Cacna1g gene induce pure absence epilepsy through genetic enhancement of the thalamocortical network (Ernst et al. 2009). One can infer from these findings that pure absence seizures in human beings, such as typical CAE, may originate from enhanced T-type current due to mutations or alternative transcripts of the three T-type calcium channel genes. The identification of T-type variants implicated in common complex epilepsy provides an important step for us in understanding this complex genetic disease.

7.3.1 CACNA1H Is a Susceptibility Gene of Idiopathic Epilepsy

7.3.1.1 Variants of CACNA1H Gene Detected in Patients with CAE

We have conducted direct sequencing of exons 3–35 and the exon–intron boundaries of the Cacna1h gene in 118 patients with CAE of Han ethnicity recruited from Northern China. Sixty-eight variations have been detected in the Cacna1h gene, and, among the variations identified, 12 were missense mutations and found in only 14 of the 118 patients in a heterozygous state but none in the 230 unrelated controls. The identified missense mutations occurred in highly conserved residues of the T-type calcium channel gene. These mutations were further introduced into human Cav3.2 cDNA and transfected into HEK-293 cells for whole-cell patch-clamp recordings. Computer simulations predicted that some mutations favor burst firings. More experimental evidence revealed that many mutant channels are activated in response to small voltage changes, an alteration in the rate of recovery of channels from their inactivated state, or an increase in the surface expression of the channels (Chen et al. 2003a; Khosravani et al. 2004, 2005; Lu et al. 2005; Vitko et al. 2005; Peloquin et al. 2006).

7.3.1.2 Variants Detected in Other Types of Idiopathic Epilepsy

Research results from Heron et al. (2007) further support the original concept that the Cacna1h gene is a susceptibility gene in absence seizure, which is also associated with an extended spectrum of idiopathic generalized epilepsies in the Caucasian population. In their study, 240 epilepsy patients and 95 control subjects were tested. More than 100 variants were detected, including 19 novel variants involving amino acid changes in subjects with phenotypes including childhood absence seizures, juvenile absence seizures, juvenile myoclonic, and myoclonic astatic epilepsies, as well as febrile seizures and temporal lobe epilepsy. Electrophysiological analysis of 11 variants showed that 9 had altered channel properties, generally in ways that should increase calcium current (Heron et al. 2007).

7.3.1.3 Variations in CACNA1G and CACNA1I in Patients with CAE

To evaluate the *CACNA1G* (alpha 1G) gene's contribution to the pathological mechanism of common children absence epilepsy, we also sequenced all of the exons in the *CACNA1G* gene in 48 patients within a Chinese population with CAE but failed to find a link between alpha 1G and human absence epilepsy (Chen et al. 2003b). However, a recent report described several putative functional variants of the *CACNA1G* gene in patients with idiopathic generalized epilepsy (IGE). The Ala570Val variant was found in one of 123 IGE patients but was not observed in a pool of 360 healthy controls. The Ala1089Ser substitution segregated into three juvenile myoclonic epilepsy (JME) affected members of a two-generation Japanese family and in one healthy control. In addition, an Asp980Asn substitution was found in two JME patients and three control individuals (Singh et al. 2007). To date, no report has shown a link between patients with idiopathic epilepsy and mutations of *CACNA1I* (Wang et al. 2006).

7.4 Alternative Transcripts of Cacna1h Gene May Be Major Factors Implicated in the Formation of Complex Absence Seizures

Of the three T-type calcium channels, alpha 1G subunits are the predominant source of T-type channels in thalamocortical relay (TC) nuclei; 1H and 1I subunits are localized to the thalamic reticular nucleus (RT), and all three forms are found in distinct but overlapping layers of the neocortex (Talley et al. 1999). The characteristic of age-dependent remission in 70 % of the patients with CAE suggests that gene dynamic expression is involved in the mechanism of CAE. Evidence has shown that subtle modifications in T-type channel gating have profound consequences for synaptically evoked firing dynamics in native neurons (Tscherter et al. 2011). All three T-type calcium channel genes have multiple transcripts, and these isoforms likely have different biophysical properties, accounting for the complex genetic characteristics of absence seizure. For example, both alpha 1H (Cav3.2) and alpha 1I (Cav3.3) transcripts are expressed in RT neurons; it has been established that Cav3.3-mediated currents generate bursts that are most closely correlated with typical RT bursts (Cain and Snutch 2010). However, Cav3.2 channels are also expressed in RT an observation consistent with the presence of Cav3.2-like currents in these neurons. The fast activation kinetics and the lower threshold of activation of the Cav3.2 channels may play a critical role in the initiation of bursts as well as the appearance of important physiological contributions that become apparent during a series of multiple bursts (Cain and Snutch 2010). Individuals may have diverse T-type calcium isoforms; if some of these unique isoforms increase T-type current, they may facilitate aberrant cortical synchronization and possibly initiate absence seizures. Therefore, the inherited complexity of multiple variants of the T-type calcium gene may be expressed through different T-type calcium channel genes to initiate CAE development.

The Cacna1h gene (Cav3.2) is located on chromosome 16p13.3 and expressed in the thalamic reticular nucleus. It is often alternatively spliced and generates a family of variant transcripts. Results on the website of National Center for Biotechnology Information (NCBI)'s AceView, which analyzed 115 GenBank accessions from 104 alpha 1H cDNA clones, illustrate that the human *CACNA1H* gene contains 38 distinct introns. Transcription produces 10 different mRNAs, 7 alternatively spliced variants, and 3 unspliced forms. There are 3 probable alternative promoters, 2 non-overlapping alternative last exons, and 2 validated alternative polyadenylation sites. The mRNAs appear to differ by truncation at the 5′ and 3′ ends, the presence or absence of a cassette exon, and overlapping exons with different boundaries. Structural changes created by missense mutations may differentially affect the activity of alternative gene products, whereas missense, silent, and noncoding mutations may alter developmental regulation of splice-variant expression. Zhang's research (Zhang et al. 2002, 2004) illustrates that the Cacan1h gene is alternatively spliced at 12–14 sites and is capable of generating both functional and nonfunctional transcripts. Biophysical profiles of different alternative Cav3.2 forms reveal variations in kinetics and steady-state gating parameters, most likely related to altered membrane firing. Zhang et al. (2002, 2004) further examined mutations of Cacna1h gene, such as C456S, D1463N, and A1765A, which appear to be unique in Chinese CAE patients but elicit minimal or no changes in CaV3.2 function, in relationship to candidate exonic splicing enhancer sequences. These missense and silent mutations appear to create or change the regulatory specificity of exonic splicing enhancer sequences that control splicing regulation. The Cacna1h gene is highly variable. According to our data (Chen et al. 2003a) and analysis of NCBI SNP data, the average SNP density of exons in the Cacna1h gene is almost 1 SNP for every 48 base pairs. The average SNP density of introns in the Cacna1h gene is almost one SNP for every 64 base pairs. It is 40 times higher in exons and 20 times higher in introns than the average genomic density. Many of the SNP's allele frequency rates are above 10 %. A low/high SNP frequency rate involves a complicated profile, where everyone can have their personal unique SNP variant system. Variants of the Cacna1h gene may destroy, create, or alter the regulatory specificity of predicted exonic splicing enhancer sequences that control splicing regulation. For example, the GAERS animal model, which contains a point mutation of the Cacna1h gene (R1584P), contain splice-variant-specific effects, requiring the presence of Exon25 to produce significantly faster recovery from channel inactivation and enhanced charge transference during high-frequency bursts (Powell et al. 2009). More research is required to clarify how variants of the Cacna1h gene affect its transcripts and how these alternative transcripts relate to the expression of CAE.

One of the main precipitating factors provoking CAE seizures is hyperventilation. As a result, carbon dioxide levels decrease in the blood, which causes the body's pH to become more alkaline. Interestingly, reducing agents such as dithiothreitol (DTT) selectively enhance native T-type current in reticular thalamic (RT) neurons and recombinant Cav3.2 (alpha 1H) current, but not native and recombinant Cav3.1 (alpha 1G) and Cav3.3 (alpha 1I)-based currents. Research shows that only Cav3.2 is unregulated by reducing agents and that reducing agents increase burst firing of RT neurons in wild-type rats and mice, but not Cav3.2 knockout mice; it also revealed an important role for the Cav3.2 isoform in thalamic signaling (Joksovic et al. 2006). More work is required to identify a possible connection between hyperventilation and transient elevation in T-type current, such as alpha 1H's T-type current in the future.

Conclusions and Future Directions

Absence seizures delineate bilaterally synchronous burst firing of an ensemble of reciprocally connected neuronal populations located in the thalamus and neocortex. Extensive research has demonstrated that both thalamic and cortical circuits participate in spike–wave activity. The concept that the thalamocortical circuit can become a widespread 3 Hz oscillator that disrupts either thalamic or cortical portions of the circuit, thereby contributing to seizures, is widely accepted by scientists. Evidence of a link between T-type calcium channel genes and absence seizures has also received considerable support since the introduction of absence seizure animal research models or examination of variations in T-type calcium channel genes in patients with epilepsy. Abnormal components, by enhancing this T-type current-associated circuit, may initiate absence seizures. Mutations of the Cacna1a gene and its ancillary calcium current occur in patients and mice with complex phenotypes. However, overexpression of Cacna1g using a transgenic mouse model results in spike–wave discharges and "pure" absence seizures. This suggests that increasing some alternative transcripts of T-type calcium channel genes may be enough to initiate pure type absence seizure. Extensive alternative splice isoform profiling of humans containing Cacna1h, Cacna1g, and Cacna1i genes demonstrates highly diversified T-type channel mRNA transcript populations with distinct electrophysiology properties (Cain and Snutch 2010). The genetic complexity of CAE may arise from different genes or one gene with differential alleles that work together to lead to abnormal enhancement of T-type current. As our data illustrate, the Cacna1h gene has a high degree of variation in each individual. Variants of this T-type calcium channel genes may alter channel gate character, increase gene expression levels, or create expressive alternative transcripts. These complexities result in elevations in T-type current, which in turn enhances the bursting activity of thalamic neurons, thereby increasing the ability to recruit network oscillatory actions in the thalamocortical circuits and initiate absence seizures. In the near future, more work is required on the relationship between variants and alternative transcripts in the role of absence seizure. The characteristic remission of absence seizures with age suggests that gene expression is involved in the mechanism of CAE. Valuable information can be gleaned from research examining the dynamic epigenetic process of T-type calcium channels.

Acknowledgements We thank Peggy Mankin and Hongyi Chen for manuscript preparation. This work was funded by the Pediatrics Department fund, Children's Hospital of Illinois.

References

Astori S, Wimmer RD, Prosser HM, Corti C, Corsi M, Liaudet N, Volterra A, Franken P, Adelman JP, Luthi A (2011) The Ca(V)3.3 calcium channel is the major sleep spindle pacemaker in thalamus. Proc Natl Acad Sci USA 108:13823–13828

Barton ME, Eberle EL, Shannon HE (2005) The antihyperalgesic effects of the T-type calcium channel blockers ethosuximide, trimethadione, and mibefradil. Eur J Pharmacol 521:79–85

Becker AJ, Pitsch J, Sochivko D, Opitz T, Staniek M, Chen CC, Campbell KP, Schoch S, Yaari Y, Beck H (2008) Transcriptional upregulation of Cav3.2 mediates epileptogenesis in the pilocarpine model of epilepsy. J Neurosci 28:13341–13353

Broicher T, Seidenbecher T, Meuth P, Munsch T, Meuth SG, Kanyshkova T, Pape HC, Budde T (2007) T-current related effects of antiepileptic drugs and a Ca^{2+} channel antagonist on thalamic relay and local circuit interneurons in a rat model of absence epilepsy. Neuropharmacology 53:431–446

Burgess DL, Noebels JL (1999a) Single gene defects in mice: the role of voltage-dependent calcium channels in absence models. Epilepsy Res 36:111–122

Burgess DL, Noebels JL (1999b) Voltage-dependent calcium channel mutations in neurological disease. Ann N Y Acad Sci 868:199–212

Cain SM, Snutch TP (2010) Contributions of T-type calcium channel isoforms to neuronal firing. Channels 4:475–482

Cain SM, Snutch TP (2013) T-type calcium channels in burst-firing, network synchrony, and epilepsy. Biochim Biophys Acta 1828:1572–1578

Chang BS, Lowenstein DH (2003) Epilepsy. N Engl J Med 349:1257–1266

Chen Y, Lu J, Pan H, Zhang Y, Wu H, Xu K, Liu X, Jiang Y, Bao X, Yao Z, Ding K, Lo WH, Qiang B, Chan P, Shen Y, Wu X (2003a) Association between genetic variation of CACNA1H and childhood absence epilepsy. Ann Neurol 54:239–243

Chen Y, Lu J, Zhang Y, Pan H, Wu H, Xu K, Liu X, Jiang Y, Bao X, Zhou J, Liu W, Shi G, Shen Y, Wu X (2003b) T-type calcium channel gene alpha (1G) is not associated with childhood absence epilepsy in the Chinese Han population. Neurosci Lett 341:29–32

Cribbs LL, Lee JH, Yang J, Satin J, Zhang Y, Daud A, Barclay J, Williamson MP, Fox M, Rees M, Perez-Reyes E (1998) Cloning and characterization of alpha1H from human heart, a member of the T-type Ca^{2+} channel gene family. Circ Res 83:103–109

Crunelli V, Leresche N (2002) Childhood absence epilepsy: genes, channels, neurons and networks. Nat Rev Neurosci 3:371–382

Erisir A, Van Horn SC, Bickford ME, Sherman SM (1997a) Immunocytochemistry and distribution of parabrachial terminals in the lateral geniculate nucleus of the cat: a comparison with corticogeniculate terminals. J Comp Neurol 377:535–549

Erisir A, Van Horn SC, Sherman SM (1997b) Relative numbers of cortical and brainstem inputs to the lateral geniculate nucleus. Proc Natl Acad Sci USA 94:1517–1520

Ernst WL, Zhang Y, Yoo JW, Ernst SJ, Noebels JL (2009) Genetic enhancement of thalamocortical network activity by elevating alpha 1G-mediated low-voltage-activated calcium current induces pure absence epilepsy. J Neurosci 29:1615–1625

Fisher RS, van Emde BW, Blume W, Elger C, Genton P, Lee P, Engel J Jr (2005) Epileptic seizures and epilepsy: definitions proposed by the International League Against Epilepsy (ILAE) and the International Bureau for Epilepsy (IBE). Epilepsia 46:470–472

Fletcher CF, Frankel WN (1999) Ataxic mouse mutants and molecular mechanisms of absence epilepsy. Hum Mol Genet 8:1907–1912

Fletcher CF, Lutz CM, O'Sullivan TN, Shaughnessy JD Jr, Hawkes R, Frankel WN, Copeland NG, Jenkins NA (1996) Absence epilepsy in tottering mutant mice is associated with calcium channel defects. Cell 87:607–617

Fuentealba P, Steriade M (2005) The reticular nucleus revisited: intrinsic and network properties of a thalamic pacemaker. Prog Neurobiol 75:125–141

Futatsugi Y, Riviello JJ Jr (1998) Mechanisms of generalized absence epilepsy. Brain Dev 20:75–79

Heron SE, Khosravani H, Varela D, Bladen C, Williams TC, Newman MR, Scheffer IE, Berkovic SF, Mulley JC, Zamponi GW (2007) Extended spectrum of idiopathic generalized epilepsies associated with CACNA1H functional variants. Ann Neurol 62:560–568

Joksovic PM, Nelson MT, Jevtovic-Todorovic V, Patel MK, Perez-Reyes E, Campbell KP, Chen CC, Todorovic SM (2006) Cav3.2 is the major molecular substrate for redox regulation of T-type Ca^{2+} channels in the rat and mouse thalamus. J Physiol 574:415–430

Khosravani H, Zamponi GW (2006) Voltage-gated calcium channels and idiopathic generalized epilepsies. Physiol Rev 86:941–966

Khosravani H, Altier C, Simms B, Hamming KS, Snutch TP, Mezeyova J, McRory JE, Zamponi GW (2004) Gating effects of mutations in the Cav3.2 T-type calcium channel associated with childhood absence epilepsy. J Biol Chem 279:9681–9684

Khosravani H, Bladen C, Parker DB, Snutch TP, McRory JE, Zamponi GW (2005) Effects of Cav3.2 channel mutations linked to idiopathic generalized epilepsy. Ann Neurol 57:745–749

Kim D, Song I, Keum S, Lee T, Jeong MJ, Kim SS, McEnery MW, Shin HS (2001) Lack of the burst firing of thalamocortical relay neurons and resistance to absence seizures in mice lacking alpha(1G) T-type Calcium channels. Neuron 31:35–45

Kostopoulos GK (2001) Involvement of the thalamocortical system in epileptic loss of consciousness. Epilepsia 42(Suppl 3):13–19

Lee JH, Daud AN, Cribbs LL, Lacerda AE, Pereverzev A, Klockner U, Schneider T, Perez-Reyes E (1999) Cloning and expression of a novel member of the low voltage-activated T-type calcium channel family. J Neurosci 19:1912–1921

Leresche N, Parri HR, Erdemli G, Guyon A, Turner JP, Williams SR, Asprodini E, Crunelli V (1998) On the action of the anti-absence drug ethosuximide in the rat and cat thalamus. J Neurosci 18:4842–4853

Liao YF, Tsai ML, Chen CC, Yen CT (2011) Involvement of the Cav3.2 T-type calcium channel in thalamic neuron discharge patterns. Mol Pain 7:43

Liu XB, Honda CN, Jones EG (1995) Distribution of four types of synapse on physiologically identified relay neurons in the ventral posterior thalamic nucleus of the cat. J Comp Neurol 352:69–91

Lowenstein D, Messing R (2007) Epilepsy genetics: yet more exciting news. Ann Neurol 62:549–550

Lu Y, Wang X (2009) Genes associated with idiopathic epilepsies: a current overview. Neurol Res 31:135–143

Lu JJ, Zhang YH, Chen YC, Pan H, Wang JL, Zhang L, Wu HS, Xu KM, Liu XY, Tao LD, Shen Y, Wu XR (2005) T-type calcium channel gene-CACNA1H is a susceptibility gene to childhood absence epilepsy. Zhonghua Er Ke Za Zhi 43:133–136

Marescaux C, Vergnes M (1995) Genetic absence epilepsy in rats from Strasbourg (GAERS). Ital J Neurol Sci 16:113–118

McCormick DA, Bal T (1997) Sleep and arousal: thalamocortical mechanisms. Annu Rev Neurosci 20:185–215

Mulley JC, Scheffer IE, Harkin LA, Berkovic SF, Dibbens LM (2005) Susceptibility genes for complex epilepsy. Hum Mol Genet 14:R243–R249

Panayiotopoulos CP, Obeid T, Waheed G (1989) Differentiation of typical absence seizures in epileptic syndromes. A video EEG study of 224 seizures in 20 patients. Brain 112:1039–1056

Peloquin JB, Khosravani H, Barr W, Bladen C, Evans R, Mezeyova J, Parker D, Snutch TP, McRory JE, Zamponi GW (2006) Functional analysis of Ca3.2 T-type calcium channel mutations linked to childhood absence epilepsy. Epilepsia 47:655–658

Perez-Reyes E (2003) Molecular physiology of low-voltage-activated T-type calcium channels. Physiol Rev 83:117–161

Perez-Reyes E, Cribbs LL, Daud A, Lacerda AE, Barclay J, Williamson MP, Fox M, Rees M, Lee JH (1998) Molecular characterization of a neuronal low-voltage-activated T-type calcium channel. Nature 391:896–900

Perez-Reyes E, Lee JH, Cribbs LL (1999) Molecular characterization of two members of the T-type calcium channel family. Ann N Y Acad Sci 868:131–143

Pinault D, O'Brien TJ (2005) Cellular and network mechanisms of genetically-determined absence seizures. Thalamus Relat Syst 3:181–203

Porter RJ (1993) The absence epilepsies. Epilepsia 34(Suppl 3):S42–S48

Powell KL, Cain SM, Ng C, Sirdesai S, David LS, Kyi M, Garcia E, Tyson JR, Reid CA, Bahlo M, Foote SJ, Snutch TP, O'Brien TJ (2009) A Cav3.2 T-type calcium channel point mutation has splice-variant-specific effects on function and segregates with seizure expression in a polygenic rat model of absence epilepsy. J Neurosci 29:371–380

Rajakulendran S, Graves TD, Labrum RW, Kotzadimitriou D, Eunson L, Davis MB, Davies R, Wood NW, Kullmann DM, Hanna MG, Schorge S (2010) Genetic and functional characterisation of the P/Q calcium channel in episodic ataxia with epilepsy. J Physiol 588:1905–1913

Rajakulendran S, Kaski D, Hanna MG (2012) Neuronal P/Q-type calcium channel dysfunction in inherited disorders of the CNS. Nat Rev Neurol 8:86–96

Ramcharan EJ, Gnadt JW, Sherman SM (2000) Burst and tonic firing in thalamic cells of unanesthetized, behaving monkeys. Vis Neurosci 17:55–62

Sherman SM (2001) Tonic and burst firing: dual modes of thalamocortical relay. Trends Neurosci 24:122–126

Singh B, Monteil A, Bidaud I, Sugimoto Y, Suzuki T, Hamano S, Oguni H, Osawa M, Alonso ME, Delgado-Escueta AV, Inoue Y, Yasui-Furukori N, Kaneko S, Lory P, Yamakawa K (2007) Mutational analysis of CACNA1G in idiopathic generalized epilepsy. Hum Mutat 28:524–525

Song I, Kim D, Choi S, Sun M, Kim Y, Shin HS (2004) Role of the alpha1G T-type calcium channel in spontaneous absence seizures in mutant mice. J Neurosci 24:5249–5257

Steriade M (1974) Interneuronal epileptic discharges related to spike-and-wave cortical seizures in behaving monkeys. Electroencephalogr Clin Neurophysiol 37:247–263

Steriade M, Contreras D (1995) Relations between cortical and thalamic cellular events during transition from sleep patterns to paroxysmal activity. J Neurosci 15:623–642

Steriade M, Deschenes M, Domich L, Mulle C (1985) Abolition of spindle oscillations in thalamic neurons disconnected from nucleus reticularis thalami. J Neurophysiol 54:1473–1497

Steriade M, Domich L, Oakson G (1986) Reticularis thalami neurons revisited: activity changes during shifts in states of vigilance. J Neurosci 6:68–81

Steriade M, Parent A, Pare D, Smith Y (1987) Cholinergic and non-cholinergic neurons of cat basal forebrain project to reticular and mediodorsal thalamic nuclei. Brain Res 408:372–376

Steriade M, McCormick DA, Sejnowski TJ (1993) Thalamocortical oscillations in the sleeping and aroused brain. Science 262:679–685

Talley EM, Cribbs LL, Lee JH, Daud A, Perez-Reyes E, Bayliss DA (1999) Differential distribution of three members of a gene family encoding low voltage-activated (T-type) calcium channels. J Neurosci 19(6):1895–1911

Talley EM, Solorzano G, Depaulis A, Perez-Reyes E, Bayliss DA (2000) Low-voltage-activated calcium channel subunit expression in a genetic model of absence epilepsy in the rat. Brain Res Mol Brain Res 75:159–165

Timofeev I, Grenier F, Steriade M (1998) Spike-wave complexes and fast components of cortically generated seizures. IV Paroxysmal fast runs in cortical and thalamic neurons. J Neurophysiol 80:1495–1513

Tringham E, Powell KL, Cain SM, Kuplast K, Mezeyova J, Weerapura M, Eduljee C, Jiang X, Smith P, Morrison JL, Jones NC, Braine E, Rind G, Fee-Maki M, Parker D, Pajouhesh H, Parmar M, O'Brien TJ, Snutch TP (2012) T-type calcium channel blockers that attenuate thalamic burst firing and suppress absence seizures. Sci Transl Med 4:121ra19

Tsakiridou E, Bertollini L, de Curtis M, Avanzini G, Pape HC (1995) Selective increase in T-type calcium conductance of reticular thalamic neurons in a rat model of absence epilepsy. J Neurosci 15:3110–3117

Tscherter A, David F, Ivanova T, Deleuze C, Renger JJ, Uebele VN, Shin HS, Bal T, Leresche N, Lambert RC (2011) Minimal alterations in T-type calcium channel gating markedly modify physiological firing dynamics. J Physiol 589:1707–1724

Vitko I, Chen Y, Arias JM, Shen Y, Wu XR, Perez-Reyes E (2005) Functional characterization and neuronal modeling of the effects of childhood absence epilepsy variants of CACNA1H, a T-type calcium channel. J Neurosci 25:4844–4855

Wang J, Zhang Y, Liang J, Pan H, Wu H, Xu K, Liu X, Jiang Y, Shen Y, Wu X (2006) CACNA1I is not associated with childhood absence epilepsy in the Chinese Han population. Pediatr Neurol 35:187–190

Westenbroek RE, Sakurai T, Elliott EM, Hell JW, Starr TV, Snutch TP, Catterall WA (1995) Immunochemical identification and subcellular distribution of the alpha 1A subunits of brain calcium channels. J Neurosci 15:6403–6418

Zhang Y, Mori M, Burgess DL, Noebels JL (2002) Mutations in high-voltage-activated calcium channel genes stimulates low-voltage-activated currents in mouse thalamic relay neurons. J Neurosci 22:6362–6371

Zhang Y, Vilaythong AP, Yoshor D, Noebels JL (2004) Elevated thalamic low-voltage-activated currents precede the onset of absence epilepsy in the SNAP25-deficient mouse mutant coloboma. J Neurosci 24:5239–5248

Regulation of T-Type Ca^{2+} Channels in Cancer Cell Cycle

8

Jonathan E. Pottle and Lloyd S. Gray

Abstract

Calcium signaling plays an important role in the regulation of cell cycling. Cells passing from G1 to S require both extracellular calcium and functional plasma membrane calcium channels in order to activate various downstream enzymes, such as ribonucleotide reductase, thymidine kinase, thymidylate synthase, and DNA polymerase. Fast proliferating cancer cells contain higher levels of basal $[Ca^{2+}]_i$ than normal epithelial cells. Indeed, in order to compensate for the increased requirement for Ca^{2+} during periods of high rates of cell proliferation, cancer cells frequently express more T-type Ca^{2+} channels to provide an extra source for Ca^{2+} influx. Thus, blockade of T-type Ca^{2+} channels in cancer cells can synchronize the pace of cell cycling. Using such an "interlaced" therapeutic approach could enhance the effectiveness of chemotherapy. For example, if T-type Ca^{2+} channel blockers stopped proliferating cells near the G1/S checkpoint of the cell cycle, then at the cessation of calcium channel blocker treatment a relatively large population of tumor cells would enter S phase and be susceptible to S phase-specific chemotherapy.

8.1 Calcium and Cell Proliferation

Calcium is one of the most essential and ubiquitous signal transduction elements in eukaryotic cellular functions, including action potentials, secretion, and motility (Ferrante and Triggle 1990; Zamponi and Snutch 1998; Zhuang et al. 2000; Jones

J.E. Pottle (✉)
Department of Physiology, Tulane University Health Sciences Center, New Orleans, LA 70112, USA
e-mail: jonathanpottle@gmail.com

L.S. Gray
Tau Therapeutics, LLC, Charlottesville, VA 22902, USA

S.W. Schaffer, M. Li (eds.), *T-type Calcium Channels in Basic and Clinical Science*,
DOI 10.1007/978-3-7091-1413-1_8

1998). Precise control of intracellular calcium levels is also crucial in cell cycle progression and regulation (Ciapa et al. 1994; reviewed in Capiod 2011). Indeed, unchecked increases in intracellular calcium or dysfunctional calcium signaling can start a series of events that end in cell death (Choi 1987). Therefore, exquisite regulation of calcium signaling is vital to ensure cell growth and survival. Proper stimulation can lead to increases in cytosolic calcium into the micromolar concentration range via intracellular calcium store release through ryanodine receptors (RyR) and/or inositol trisphosphate receptors (IP3R) and calcium influx through plasma membrane voltage- and/or ligand-gated calcium channels. Calcium flows into the cytosol down the large electrochemical gradient established between the extracellular medium (1.3–2 mM) and the cytosol (<100 nM).

Calcium signaling plays an important role in the regulation of cell cycling. The cell cycle is descriptively divided into four sequential phases: G1, S, G2, and M. The S phase is characterized by DNA synthesis, while mitosis is the defining event of M phase. The transitions from G1 to S and G2 to M are characterized by restriction points through which a cell must pass to continue proliferation. If a cell is not of the proper disposition to pass through the G1/S checkpoint, it exits the cell cycle into a G0 phase and either differentiates or otherwise terminates. Passage through the restriction points requires a number of signals, such as appropriate calcium levels. For instance, introduction of exogenous calcium initiates mitosis in fertilized eggs (Trump and Berezesky 1995). Transitions from G1 to S and G2 to M were observed to be preceded by transient increases in cytosolic calcium concentration, phenomena that are dependent on physiological extracellular calcium concentrations (Steinhardt and Alderton 1988). Cells passing from G1 to S require extracellular calcium and functional plasma membrane calcium channels in order to activate various downstream enzymes, such as ribonucleotide reductase, thymidine kinase, thymidylate synthase, and DNA polymerase. Microtubule rearrangement and microfilament contraction, crucial to progression to M phase and mitosis, are dependent on enzymes that are activated by "calcium flashes." Progression through the cell cycle in fertilized eggs has been blocked with calcium chelators (Zucker and Steinhardt 1978). G0/G1 cell cycle arrest has been induced in growth factor-induced human umbilical arterial endothelial cells via application of various calcium channel blockers (Zeitler et al. 1997). Immediate early genes such as *c-fos*, which induce cells in G0 to reenter the active cell cycle, have been associated with calcium signaling; activation of these genes is a hallmark of rapidly proliferating cells.

Given appropriate conditions, cells can "commit suicide" through a genetically programmed process called apoptosis. Apoptosis aids in the development of organs and the maintenance of healthy cell populations. Calcium signaling is crucial to the initiation and progression of apoptosis through a calcium surge and interaction with the nuclear membrane calcium-activated endonuclease, an enzyme that fragments chromatin. Calcium has other important roles in cell proliferation via more indirect signaling, involving signal transduction pathways such as protein kinase C, G-proteins, m-calpain, various MAP kinases, calmodulin, and phospholipase A2 (Ariyoshi et al. 1998; Akagi et al. 1999). An example of this is seen in the action of phospholipase C. A rise in intracellular calcium concentration activates phospholipase C, which then catalyzes the hydrolysis of membrane inositol-containing lipids to produce IP3 and diacylglycerol. A downstream messenger in signaling cascades descending from G-protein or tyrosine kinase-linked membrane receptors (Berridge 1993), IP3 binds to IP3R in the endoplasmic reticulum membrane to induce release of calcium from the ER lumen into the cytosol. This calcium-induced calcium release can be amplified and can support further signaling to trigger events such as DNA synthesis and cell division (Lechleiter and Clapham 1992).

There are many membrane-associated molecules that are responsive to free calcium ions, e.g., members of the phospholipase C superfamily (Meldrum et al. 1991), copines (Creutz et al. 1998), and protein kinase Cα (Micol et al. 1999), to name but a few. Many calcium signals involve calcium-sensor proteins. Calmodulin is an important calcium-binding adapter protein that functions as a calcium concentration monitor during the cell cycle. Calmodulin was expressed at high levels during DNA synthesis and mitosis, and administration of antibodies for calmodulin prevented DNA synthesis (Reddy et al. 1992). The extracellular calcium ion concentration sensing receptor (CaR) and the calcium-binding regulatory protein calbindin, which is related to calmodulin, both function in calcium sensing mechanisms. CaR participates in sensing and reacting to fluctuations in extracellular calcium concentration; increased extracellular calcium concentration leads to CaR interacting with phospholipase C via G-proteins, resulting in the formation of IP3 and diacylglycerol (Kifor et al. 1997). Calbindin regulates the increase in intracellular calcium caused by ER store release. Calbindin has also been shown to interact with L-type calcium channels in pancreatic beta cells (Parkash et al. 2002). CaR/calbindin co-localization in MCF-7 breast cancer cells has been observed (Parkash et al. 2004). MCF-7 cell metastasis has been promoted by increased parathyroid hormone-related protein (PTHrP), which is stimulated by activated CaR (Sanders et al. 2000). Additionally, increased intracellular calcium concentration regulated cancer cell transendothelial migration in an in vitro model (Lewalle et al. 1998). These examples illustrate the importance and diversity of calcium signaling in cell growth.

Cancer cells demonstrate altered calcium signaling during proliferation (Whitfield 1992). Normal liver cells require millimolar extracellular calcium concentrations in order to successfully synthesize DNA and proceed through the cell cycle, while cancerous liver cells are able to maintain the molecular effectors of chromosomal replication and spontaneously proliferate even in largely calcium-deficient extracellular fluid. These liver cancer cells are also unresponsive to other usual controllers of proliferation, such as TGF-β, and can even produce their own growth factors that reduce the need for extracellular calcium (Whitfield 1992). Tumor suppressor-deficient colon carcinomas have been shown to display widely deviant calcium signaling pathways and insensitivity to normal calcium-dependent restrictions, thanks to overexpression of calcium-binding signal proteins (Kifor et al. 1997). Liver cancer cells have been demonstrated to overproduce calmodulin (Whitfield 1992), a multifunctional calcium-binding effector that can stimulate proliferation [e.g., in mouse fibroblasts (Rasmussen and Means 1989)]. Human

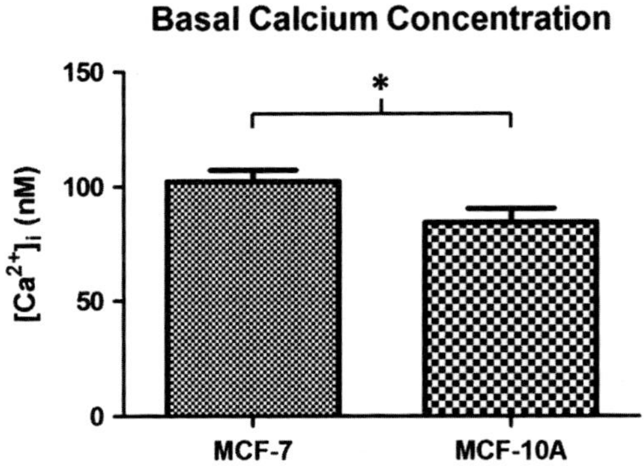

Fig. 8.1 Basal calcium concentration in MCF-7 and MCF-10A. The tumorigenic MCF-7 cell line was found to have significantly higher basal calcium concentration than that found in the non-tumorigenic MCF-10A cell line (Pottle et al. 2013)

myeloid leukemia cells release calmodulin into the extracellular fluid, where it spurs DNA synthesis and cell proliferation (Whitfield 1992). Cancer cells have also been found to produce macrophage-derived growth factor (Yin et al. 2006) and calcium-binding parvalbumin protein oncomodulin, which can stimulate DNA replication (Whitfield 1992). Oncomodulin normally participates in development (Whitfield 1992) and axon regeneration (Yin et al. 2006), but the gene is activated in cancers of the liver, colon, and skin (Whitfield 1992). Examination of primary human liver tumors revealed that high-grade tumors lack E-cadherin expression, while lower grade carcinomas and normal tissues retain expression (Shimoyama and Hirohashi 1991). E-cadherin is a calcium-binding cell adhesion molecule, and loss of its expression allows high-grade malignant cells to migrate out from the tumor site, and indeed the higher grade carcinomas demonstrate looser cell–cell connections. Such cells with dysfunctional calcium signaling cannot be terminated via typically effective calcium signals. We found that MCF-7 breast cancer cells maintain a basal calcium concentration higher than that of non-tumor MCF-10A cells (Fig. 8.1). The roles of calcium and downstream effectors in cancer cell cycle progression must be better understood if a better picture of cancer cell biology is to be painted.

8.2 T-Type Calcium Channels in Cell Proliferation

The role of low-voltage-activated T-type calcium channels in cell proliferation has been established, though the mechanistic details remain largely elusive. T-type calcium current was observed in primary cultured rat aortic smooth muscle cells during the G1 and S phases of the cell cycle but was found to be significantly reduced or altogether absent in other cell cycle phases (Kuga et al. 1996; Guo et al. 1998; Wang et al. 2002; Li et al. 2005). Proliferating cells express T-type calcium channels at a density higher than that of other calcium channels (Kuga et al. 1996). Evidence indicates that T-type calcium channels are commonly found in the rapidly growing cells of developing organs, while they are largely absent in cells that are not proliferating and normal epithelial cells, outside of pathological cell growth. The proliferation of human pulmonary artery myocytes is reported to be regulated by T-type calcium channels (Rodman et al. 2005). Richard et al. (1992) found that T-type calcium current was detected in cultured smooth muscle cells during proliferation, but that current decreased as the cell monolayers approached confluence or cells contacted one another. Wolfe et al. (2003) found that the membrane-bound G-protein-coupled receptor β-γ subunit, specifically the β2-γ2 combination, plays a role in inhibiting T-type calcium current carried by α1H channels; however, they found no role for β-γ in the regulation of T-type calcium channel membrane expression and no effects on the current flowing through α1G channels. Expression of T-type calcium channel mRNA was found to be stimulated by angiotensin II, through an angiotensin receptor 1-activated MEK1/2-mediated pathway, in isolated newborn rat cardiomyocytes (Ferron et al. 2003). The mitogen-activated growth pathways mediated by protein kinases such as MEK1/2 are diverse and involve massive amounts of cross talk and integration with other signaling pathways, so it is not surprising that such mechanisms might be involved in the expression of proliferation-regulating T-type calcium channels. T-type calcium channels are also present in vascular endothelial cells, where they regulate endothelial cell intracellular calcium concentration and cell migration via an angiotensin II-dependent pathway involving the AT1 receptor (Martini et al. 2010). The relationship between T-type calcium channels and the renin–angiotensin system should prove to be a complex one.

Freshly dissociated neonatal rat ventricular myocytes possess T-type calcium currents while they are proliferating, but those currents disappear in cells cultured for longer than 3 days (Gomez et al. 1994). Mouse preadipocytes express T-type calcium channels, mostly of the α1G isoform, prior to differentiation; the T-type calcium channel blocker mibefradil and its derivative NNC-55-0396 inhibit proliferation, while NNC-55-0396 and siRNA targeting α1G mRNA prevent cell cycle entry and progression (Oguri et al. 2010). The α1H isoform has been found to be integral to the proliferation of glomerular mesangial cells, which aberrantly proliferate in diseases such as glomerulosclerosis (Mulgrew et al. 2009). Pulmonary arterial smooth muscle cells that have been stimulated to grow with IGF-1 have been observed to display upregulated α1G expression that is related to cyclin D activation (Pluteanu and Cribbs 2011). Zinc transporter-1, a molecule that protects cells from zinc toxicity, increases α1G and α1H currents as well as ERK phosphorylation (Mor et al. 2012).

T-type calcium channels are but one element in a complex signaling system controlling proliferation, among other cellular functions. As befits their status as current conduits, T-type calcium channels are found at high expression levels in the dendrites of many neurons. Hildebrand et al. (2009) found that α1G T-type calcium channels co-localize with and are potentiated by the metabotropic glutamate receptor mGluR1, a participant in synaptic signaling. mGluR1 is a G-protein-coupled receptor, a superfamily of proteins that includes several members that can influence cell proliferation. The possibility that T-type calcium channels act in concert with other such molecules to control cell growth is enticing. In a possible counterexample to the general findings described above, illustrating the complexity of calcium signaling, T-type calcium channel blockers were shown to have protective effects on neurons that express T-type calcium channels, although the dominant isoform present (α1H) apparently is not the relevant target (Wildburger et al. 2009).

8.3 T-Type Calcium Channels in Cancer Cell Proliferation

The importance of calcium to cancer growth has a number of supporting factors. Breast cancers are known to deposit calcium in mammary ducts; such deposits are diagnostically useful features that can be detected radiographically during mammographic screenings (Galkin et al. 1977). Bone is a trophic site for breast cancer metastasis (Percival et al. 1985), and bone resorption makes abundant extracellular calcium available (Silver et al. 1998). However, cancer cells often are not as sensitive to low extracellular calcium concentration as normal cells are, and they do not respond to calcium signals that should trigger differentiation or apoptosis (Whitfield 1992). Cancer cells may employ aberrant calcium influx and management mechanisms in order to have enough calcium ions available temporally and spatially to facilitate their speedy proliferation without allowing intracellular calcium concentration to rise to apoptosis-inducing levels. A cell expressing T-type calcium channels would have an extra pathway for calcium ion entry into the cytosol from the extracellular fluid and so might be expected to have a slightly elevated intracellular calcium concentration at rest, via mechanisms such as window current.

Tumor cells generally have a larger degree of glucose utilization than normal cells, as it is necessary to fuel cancerous growth. Movement of glucose from the extracellular fluid into a cell is facilitated by membrane proteins called glucose transporters. Various glucose transporters (GLUTs), usually GLUTs appropriate to a given tumor's tissue of origin, have been found to be overexpressed in cancer cells and thus facilitate increased glucose consumption (reviewed in Smith 1999; Medina and Owen 2002). The glucose transporter GLUT12 was characterized in MCF-7 cells (Rogers et al. 2002) and has been found in prostate carcinoma cell lines (Chandler et al. 2003). High glucose has been found to increase T-type calcium channel mRNA and protein expression in pancreatic β-cells (Keyser et al. 2014). A compelling avenue of investigation would be to examine the relationship between glucose and T-type calcium expression in cancer cells. Complete growth medium for MCF-7 cells includes insulin, as specified by ATCC, and GLUT12 has been shown to be an insulin-responsive glucose transporter in muscle (Stuart et al. 2009). Perhaps an examination of the effects of removal of insulin from the growth medium on T-type calcium channel expression in cancer cells would yield interesting results.

If T-type calcium channels play an important role in pathological cell proliferation, normal cells must have mechanisms to control the expression of T-type calcium channel mRNA and protein. These channels have been found to participate in such abnormal conditions as hyperglycemia-induced rat cardiomyocyte proliferation (Li et al. 2005), cardiomyopathic hamster heart (Bkaily et al. 1997; Villame et al. 2001), hypertrophied adult feline left ventricular myocytes (Nuss and Houser 1993), rat neointimal formation after vascular injury (Schmitt et al. 1995), and hypoxia-induced catecholamine release by rat carotid body chemoreceptors (Cáceres et al. 2009). Also, uncontrolled expression of the mRNA and subsequently functional protein channels could lead to tumor initiation and progression. The mRNA coding for T-type calcium channels was found in biopsied human breast cancer tissue; pathological examination of the tumor from which the specimens were taken led to a diagnosis of the tumor as malignant with estrogen receptor expression (Taylor et al. 2008b). Tumor cell expression of T-type calcium channels has been reported in a wide array of tumor types (Leuranguer et al. 1998; Wyatt et al. 1998; Assandri et al. 1999; Toyota et al. 1999; Zhuang et al. 2000; Lesouhaitier et al. 2001; Bertolesi et al. 2002; Hirooka et al. 2002; Mariot et al. 2002; Wang et al. 2002; Del Toro et al. 2003; Harkins et al. 2003; Gray et al. 2004; Huang et al. 2004a; Latour et al. 2004; Panner et al. 2005; Lu et al. 2008).

T-type calcium channel current can be blocked with various agents, such as mibefradil and pimozide (Mehrke et al. 1994; Enyeart et al. 1994). Both of these agents exert other effects: mibefradil also blocks L-type calcium channels (Mehrke et al. 1994; Protas and Robinson 2000; Wu et al. 2000), and pimozide blocks potassium currents (Kang et al. 2001). Some colon cancers have been shown to express L-type calcium channels (Wang et al. 2000), but these high-voltage-activated channels were found to promote apoptosis rather than proliferation (Zawadzki et al. 2008). Lu et al. (2008) found that the mibefradil-induced upregulation of p21 in T8 esophageal cancer cells is dependent on p53. Expression of p21 is largely regulated by the activity of p53, possibly the most prolific tumor suppressor in human cells; however, p53-independent p21 modulation occurs in MCF-7 cells (e.g., Han et al. 2012). A search for more selective blockers is under way. Huang et al. (2004b) described NNC-55-0396, a mibefradil derivative that was found to resist hydrolysis and not be transformed into a metabolite with L-type calcium channel inhibiting activities. Other blockers with high specificity for T-type calcium channels are being developed and analyzed (e.g., Gu et al. 2010). Panner et al. (2005) found that the proliferation rates of U87-MG astrocytoma cells and N1E-115 neuroblastoma cells are both reduced by mibefradil; the proliferation rates of these cell lines doubled by overexpressing T-type calcium channels. These investigators also successfully decreased growth in intracranial tumors that express T-type calcium channels, a phenomenon not seen in cancer cells lacking such expression.

The roles of T-type calcium channels and their blockers in cancer cell viability and proliferation are not without controversy. Using breast cancer as an example, Bertolesi et al. (2002) demonstrated that mibefradil can inhibit the growth of MCF-7 cells in vitro, an observation that held true in some retinoblastoma cells; those investigators also found that mibefradil can cause necrotic cell death in retinoblastoma cells, though no similar experiments were described for MCF-7 cells. Such a phenomenon could confound any results that are due to the function of T-type calcium channels. If T-type calcium channels are to be considered a specific therapeutic target in breast cancer treatment, general toxicity is a side effect to be minimized in any potential drug. We found that NNC-55-0396 and T-type calcium channel knockdown via siRNA inhibit the growth of MCF-7 cells without increasing cell death via apoptosis (Taylor et al. 2008a; Pottle et al. 2013). Ohkubo and Yamazaki (2012), however, reported functional differences between α1G and α1H in MCF-7 cell survival and growth. Those authors found that α1H has little influence on MCF-7 cell viability and proliferation. Further, they observed that overexpression of α1G actually decreased proliferation via induction of apoptosis, while pharmacological blockade with the tarantula toxin ProTx-1 (an inhibitor that is reportedly selective for α1G over α1H) promoted proliferation and α1G knockdown reduced cyclophosphamide-induced apoptosis. Correspondingly, those authors observed α1G expression in the plasma membrane of cells with apoptotic morphology; α1H expression was observed mainly in healthy cells and displayed limited overlap with α1G expression. The results of these studies are certainly caused for scrutiny, though they may not be as directly in opposition as a first glance would suggest. The findings of Taylor et al. (2008a) reported both α1G and α1H expression in proliferating MCF-7 cells, while Ohkubo and Yamazaki (2012) found little overlap in isoform expression, and α1G was observed mainly in apparently apoptotic cells. The cause of this discrepancy is unclear, though cancer cells are often characterized by unstable genomes. Calcium signaling as it relates to cell survival is a matter of balance: too much intracellular calcium or too little can lead to cell death. Differing levels of expression will yield differing results; perhaps overexpression relative to "normal" MCF-7 cell α1G expression level is enough to trigger apoptosis, while baseline expression is beneficial. Cyclophosphamide induces apoptosis by interfering with DNA synthesis. A reduction in proliferation, especially halting cell cycle progression at the calcium-sensitive G1/S checkpoint, would indeed reduce the effects of an apoptosis-inducing alkylating agent such as cyclophosphamide. The differences in isoforms may be a key factor in the differences among these studies' results. Some experiments (Bertolesi et al. 2002; Taylor et al. 2008a; Pottle et al. 2013) did not discriminate between α1G and α1H in blockade or knockdown experiments, and the complexities of the relevant signaling mechanisms may mean that the effects of targeting *both* isoforms are not simply the additive result of the effects of targeting each isoform individually. Ohkubo and Yamazaki (2012) reported distinct differences in cell cycle-related gene expression profiles resulting from knockdown of α1G and α1H, indicating that the two isoforms exert different effects on intracellular processes. The potential for interaction and cross talk of signaling pathways modulated by α1G and α1H is implied. While no single dysfunctional channel is likely to serve as the sole culprit for any broad swath of cancers (Santoni et al. 2012), the functional role for T-type calcium channels in cancer cell survival and growth requires more investigation.

T-type calcium channels have also been implicated in intracranial tumors and prostate cancer. The human astrocytoma cell line U87-MG and neuroblastoma cell line N1E-115 both display mibefradil-sensitive proliferation. Furthermore, overexpression of T-type calcium channels doubled the proliferation rate of these cells, while antisense oligonucleotides directed against T-type calcium channel mRNA reduced the proliferation rate in intracranial tumor cells by 45 % but had no effect in cancer cells that lack T-type calcium channel expression (Panner et al. 2005). Other researchers have found that mibefradil's effects on U87 cell proliferation may be independent of T-type or L-type calcium channels (Zhang et al. 2012). The human prostate cancer epithelial cell line LNCaP was found to overexpress α1H T-type calcium channel mRNA and current. As with the intracranial tumor lines, overexpression of T-type calcium channels resulted in increased proliferation, and antisense oligonucleotides reduced proliferation in these cells. The channels also play a role in LNCaP intracellular calcium regulation (Mariot et al. 2002; Wang et al. 2002). T-type calcium channels may also have a functional role in esophageal carcinoma cell proliferation, as evidenced by their sensitivity to channel inhibition (Lu et al. 2008). Both α1H and α1G have been found at abnormally high expression levels in human retinoblastoma cells, human breast cancer cells, and rat glioma cells (Bertolesi et al. 2002; Lu et al. 2008), while the gene coding for the α1H subunit is expressed in several breast cancer cell lines, e.g., estrogen receptor-positive MCF-7 and triple negative MDA-MB-231 (Asaga et al. 2006). Li et al. (2011) found that blockade of T-type calcium channels with mibefradil or NNC-55-0396 and specific knockdown of expression with siRNA targeting α1G and α1H suppressed proliferation in two ovarian cancer cell lines, with an increase in cells in the G0/G1 phase of the cell cycle. The inhibitor NNC-55-0396 also decreased ovarian cancer formation in nude mice. These results suggest the possibility that T-type calcium channel blockers may have some beneficial effect in the oncology clinic.

8.4 Targeting T-Type Calcium Channels in Oncology

Chemotherapy is a common cancer treatment modality that generally relies on the susceptibility of proliferating cells to irreversible DNA damage or alteration. An issue that relates to susceptibility to chemotherapy is that of intratumoral DNA heterogeneity. Cells within a tumor grow at different rates, and cell cycle stage can vary throughout the tumor at any given time. For instance, the proportion of tumor cells in S phase in breast cancer samples has been found to vary between 2 and 32 % (Siitonen et al. 1993; Bergers et al. 1996); similar analysis in ovarian cancer samples found a range of 0-28 % (Kaern et al. 1994). An alternating or "interlaced" therapeutic approach could make chemotherapy more effective: if T-type calcium channel blockers stop proliferating cells near the G1/S checkpoint of the cell cycle, then at the cessation of channel blocker treatment, there will be a relatively large population of tumor cells entering S phase and progressing through the cell cycle. If a phase-specific chemotherapy—one that works in S phase, in this thought experiment—is then given, there will be a larger proportion of tumor cells that are susceptible to the chemotherapy than in a control heterogeneous tumor cell population. This may also allow for less chemotherapy used in any individual treatment session, because a smaller amount of the drug can kill a larger percentage of the tumor than would be the case in a naturally heterogeneous tumor. The use of such cell cycle synchronization in conjunction with chemotherapy is hardly a new idea. The use of cytostatic agents to synchronize tumor cells in order to make chemotherapy more effective has a long history in research. Agents as diverse as hydroxyurea (Volm and Wayss 1975) and microbial iron chelators (Bergeron and Ingeno 1987) have been investigated in combination with chemotherapy. Some cancers such as breast, ovarian, and prostate are responsive to hormones due to the characteristics of their tissues of origin. As such, treatments that disrupt hormone signaling are often employed to treat these cancers. Tamoxifen, an estrogen receptor antagonist, can be used to cause G1 arrest in breast cancer cells (Osborne et al. 1983). However, hormones such as estrogens have such varied and pronounced effects in the human body that the inhibition or manipulation of hormone signaling can result in numerous side effects of varying severity, including increased bone fracture due to bone loss (reviewed in Body 2011), arthralgias (reviewed in Gaillard and Stearns 2011), vision effects (reviewed in Eisner and Luoh 2011), and heart disease and stroke (reviewed in Chlebowski et al. 2006).

T-type calcium channel blockers have shown promise when used as an adjuvant modality to chemotherapy. Keir et al. (2013) demonstrated that mibefradil induces cell cycle synchronization near the G1/S checkpoint in glioblastoma cells. Further, mibefradil increased the efficacy of chemotherapy in mouse glioblastoma xenograft models. The calcium channel blocker also sensitized otherwise resistant tumors to temozolomide, underlining the importance of T-type calcium channels and intracellular calcium regulation in the growth of a type of cancer that often has poor prognosis. We obtained similar results in breast cancer cells (Taylor et al. 2008a; Pottle et al. 2013). The specific T-type calcium channel blocker NNC-55-0296 and siRNA targeting α1G and α1H mRNA both inhibited MCF-7 and MDA-MB-231 breast cancer cell growth (Taylor et al. 2008a). A treatment regimen consisting of sequentially alternating NNC-55-0396 and chemotherapy—termed "interlaced therapy" (Fig. 8.2)—proved more effective than chemotherapy alone at killing MCF-7 cells in vitro (Taylor et al. 2008a), and interlaced therapy utilizing mibefradil and chemotherapy more effectively reduced MCF-7 tumor xenograft size than chemotherapy alone in a nude mouse model (Pottle et al. 2013). The side effects of specifically inhibiting T-type calcium channels in a patient likely would be less detrimental to the quality of life than agents affecting hormonal signaling, for example. However, such potential side effects—activation of T-type calcium channels is involved in nitrous-oxide's mediation of pain signaling, for instance (Kim et al. 2009; Orestes et al. 2011)—do demand further investigation.

The efficacy of T-type calcium channel blockers in increasing the effectiveness of cancer therapies is not limited to chemotherapy. Valerie et al. (2013) investigated using T-type calcium channel inhibition in conjunction with radiation therapy. Mibefradil reduced the viability and clonogenic potential of radiation therapy-resistant glioblastoma cells compared to radiation treatment alone, and the blocker also induced apoptosis through caspase activation. Mibefradil increased the susceptibility of the resistant cells to ionizing radiation. Similar results were found for siRNA targeting T-type calcium channel mRNA, while specific inhibition of L-type calcium channels (a side effect of mibefradil) produced practically no effect.

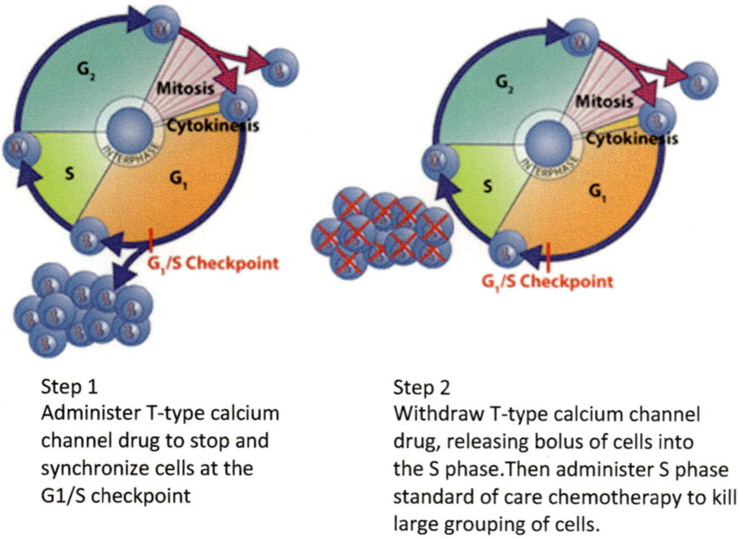

Fig. 8.2 A scheme of interlaced therapy. A regimen of alternating cytostatic and cytotoxic treatments can synchronize cancer cells in the cell cycle and improve the efficacy of chemotherapy

The results discussed above emphasize the crucial role of T-type calcium channels in the viability and proliferation of cancer cells, a role that may be exploited to increase the efficacy and reduce the side effects of cancer treatment. Inhibiting the growth of cancer cells with T-type calcium channel blockers can reduce tumor regrowth between chemotherapy or radiation therapy doses. Potentiating chemotherapy or radiation therapy can allow oncologists to use lower dosages, reducing the side effects experienced by the patient. More efficacious treatment can lead to shorter treatment schedules, realizing physical, emotional, and economic benefits.

References

Akagi K, Nagao T, Urushidani T (1999) Correlation between $Ca^{(2+)}$ oscillation and cell proliferation via CCK(B)/gastrin receptor. Biochim Biophys Acta 1452:243–253

Ariyoshi H, Okahara K, Sakon M, Kambayashi J, Kawashima S, Kawasaki T, Monden M (1998) Possible involvement of m-calpain in vascular smooth muscle cell proliferation. Arterioscler Thromb Vasc Biol 18:493–498

Asaga S, Ueda M, Jinno H, Kikuchi K, Itano O, Ikeda T, Kitajima M (2006) Identification of a new breast cancer-related gene by restriction landmark genomic scanning. Anticancer Res 26:35–42

Assandri R, Egger M, Gassmann M, Niggli E, Bauer C, Forster I, Gorlach A (1999) Erythropoeitin modulates intracellular calcium in a human neuroblastoma cell line. J Physiol 516:343–352

Bergeron RJ, Ingeno MJ (1987) Microbial iron chelator-induced cell cycle synchronization in L1210 cells: potential in combination chemotherapy. Cancer Res 47:6010–6016

Bergers E, van Diest PJ, Baak JP (1996) Tumor heterogeneity of DNA cell cycle variables in breast cancer measured by flow cytometry. J Clin Pathol 49:931–7

Berridge MJ (1993) Inositol triphosphate and calcium signaling. Nature 361:315–325

Bertolesi GE, Shi C, Elbaum L, Jollimore C, Rozenberg BS, Kelly MEM (2002) The Ca^{2+} channel antagonists mibefradil and pimozide inhibit cell growth via different cytotoxic mechanisms. Mol Pharmacol 62:210–219

Bkaily G, Sculptoreanu A, Jacques D, Jasmin G (1997) Increases of T-type Ca^{2+} current in heart cells of the cardiomyopathic hamster. Mol Cell Biochem 176:199–204

Body JJ (2011) Increased fracture rate in women with breast cancer: a review of the hidden risk. BMC Cancer 11:384

Cáceres AI, Gonzalez-Obeso E, Gonzalez C, Rocher A (2009) RT-PCR and pharmacological analysis of L- and T-type calcium channels in rat carotid body. Adv Exp Med Biol 648:105–112

Capiod T (2011) Cell proliferation, calcium influx and calcium channels. Biochimie 93:2075–2079

Chandler JD, Williams ED, Slavin JL, Best JD, Rogers S (2003) Expression and localization of GLUT1 and GLUT12 in prostate carcinoma. Cancer 97:2035–2042

Chlebowski RT, Anderson GL, Geller M, Col N (2006) Coronary heart disease and stroke with aromatase inhibitor, tamoxifen and menopausal hormone therapy use. Clin Breast Cancer 6 (Suppl 2):S58–S64

Choi DW (1987) Ionic dependence of glutamate neurotoxicity. J Neurosci 7:369–379

Ciapa B, Pesando D, Wilding M, Whitaker M (1994) Cell-cycle calcium transients driven by cyclic changes in inositol triphosphate levels. Nature 368:875–878

Creutz CE, Tomsig JL, Snyder SL, Gautier MC, Skouri F, Beisson J, Cohen J (1998) The copines, a novel class of C2 domain-containing, calcium-dependent, phospholipid-binding proteins conserved from paramecium to humans. J Biol Chem 273:1393–1402

Del Toro R, Levitsky KL, Lopez-Barneo J, Chiara MD (2003) Induction of T-type calcium channel gene expression by chronic hypoxia. J Biol Chem 278:22316–22324

Eisner A, Luoh SW (2011) Breast cancer medications and vision: effects of treatments for early-stage disease. Curr Eye Res 36:867–885

Enyeart JJ, Biagi BA, Day RN, Sheu SS, Maurer RA (1994) Blockade of low and high threshold Ca^{2+} channels by diphenylbutylpiperidine antipsychotics linked to inhibition of prolactin gene expression. J Biol Chem 265:1483–1488

Ferrante J, Triggle DJ (1990) Homologous and heterologous regulation of voltage-dependent calcium channels. Biochem Pharmacol 39:1267–1270

Ferron L, Capuano V, Ruchon Y, Deroubaix E, Coulombe A, Renaud JF (2003) Angiotensin II signaling pathways mediate expression of cardiac T-type calcium channels. Circ Res 93:1241–1248

Gaillard S, Stearns V (2011) Aromatase inhibitor-associated bone and musculoskeletal effects: new evidence defining etiology and strategies for management. Breast Cancer Res 13:205

Galkin BM, Feig SA, Patchefsky AS, Rue JW, Gamblin WJ Jr, Gomez DG, Marchant LM (1977) Ultrastructure and microanalysis of 'benign' and 'malignant' breast calcifications. Radiology 124:245–249

Gomez JP, Portreau D, Branka JE, Raymond G (1994) Developmental changes in Ca^{2+} currents from newborn rat cardiomyocytes in primary culture. Pflugers Arch 428:241–249

Gray LS, Perez-Reyes E, Gomora JC, Haverstick DM, Shattock M, McLatchie L, Harper J, Brooks G, Heady T, Macdonald TL (2004) The role of voltage gated T-type Ca2+ channel isoforms in mediating "capacitative" Ca^{2+} entry in cancer cells. Cell Calcium 36:489–497

Gu SJ, Lee JK, Pae AN, Chung HJ, Rhim H, Han SY, Min SJ, Cho YS (2010) Synthesis and biological evaluation of 1,4-diazepane derivatives as T-type calcium channel blockers. Bioorg Med Chem Lett 20:2705–2708

Guo W, Kamiya K, Kodama I, Toyama J (1998) Cell cycle-related changes in the voltage-gated Ca^{2+} currents in cultured newborn rat ventricular myocytes. J Mol Cell Cardiol 30:105–109

Han J, Kim S, Yang JH, Nam SJ, Lee JE (2012) TPA-induced p21 expression augments G2/M arrest through a p53-independent mechanism in human breast cancer cells. Oncol Rep 27:517–522

Harkins AB, Cahill AL, Powers JF, Tischler AS, Fox AP (2003) Expression of recombinant calcium channels support secretion in a mouse pheochromocytoma cell line. J Neurophysiol 90:2325–2333

Hildebrand ME, Isope P, Miyazaki T, Nakaya T, Garcia E, Feltz A, Schneider T, Hescheler J, Kano M, Sakimura K, Watanabe M, Dieudonné S, Snutch TP (2009) Functional coupling between mGluR1 and Cav3.1 T-type calcium channels contributes to parallel fiber-induced fast calcium signaling within Purkinje cell dendritic spines. J Neurosci 29:9668–9682

Hirooka K, Bertolesi GE, Kelly ME, Denovan-Wright EM, Sun X, Hamid J, Zamponi GW, Juhasz AE, Haynes LW, Barnes S (2002) T-Type calcium channel alpha1G and alpha1H subunits in human retinoblastoma cells and their loss after differentiation. J Neurophysiol 88:196–205

Huang JB, Kindzelskii AL, Clark AJ, Petty HR (2004a) Identification of channels promoting calcium spikes and waves in HT1080 tumor cells: their apparent roles in cell motility and invasion. Cancer Res 64:2482–2489

Huang L, Keyser BM, Tagmose TM, Hansen JB, Taylor JT, Zhuang H, Zhang M, Ragsdale DS, Li M (2004b) NNC 55-0396 [(1S,2S)-2-(2-(N-[(3-benzimidazol-2-yl)propyl]-N-methylamino) ethyl)-6-fluoro-1,2,3,4-tetra-hydro-1-isopropyl-2-naphtyl cyclopropanecarboxylate dihydrochloride]: a new selective inhibitor of T-type calcium channels. J Pharmacol Exp Ther 309:193–199

Jones SW (1998) Overview of voltage-dependent calcium channels. J Bioenerg Biomembr 30:299–312

Kaern J, Tropé CG, Kristensen GB, Pettersen EO (1994) Flow cytometric DNA ploidy and S-phase heterogeneity in advanced ovarian carcinoma. Cancer 73:1870–1877

Kang J, Chen XL, Rampe D (2001) The antipsychotic drugs sertindole and pimozide block erg3, a human brain $K^{(+)}$ channel. Biochem Biophys Res Commun 286:499–504

Keir ST, Friedman HS, Reardon DA, Bigner DD, Gray LA (2013) Mibefradil, a novel therapy for glioblastoma multiforme: cell cycle synchronization and interlaced therapy in a murine model. J Neurooncol 111:97–102

Keyser BM, Taylor JT, Choi S-K, Lu Y, Bhattacharjee A, Huang L, Pottle J, Matrougui K, Xu Z, Li M (2014) Role of T-type Ca^{2+} channels in basal $[Ca^{2+}]_i$ regulation and basal insulin secretion in rat islet cells. Curr Trend Endocrinol 7:35–44

Kifor O, Diaz R, Butters R, Brown EM (1997) The Ca^{2+}-sensing receptor (CaR) activates phospholipases C, A2 and D in bovine parathyroid and CaR-transfected, human embryonic kidney (HEK293) cells. J Bone Miner Res 12:715–725

Kim YH, Shim YJ, Shin YJ, Sul D, Lee E, Min BH (2009) 2,3,7,8-tetrachlorodibenzo-p-dioxin (TCDD) induces calcium influx through T-type calcium channel and enhances lysosomal exocytosis and insulin secretion in INS-1 cells. Int J Toxicol 28:151–161

Kuga T, Kobayashi S, Hiakawa Y, Kanaide H, Takeshita A (1996) Cell cycle-dependent expression of L- and T-type Ca^{2+} currents in rat aortic smooth muscle cells in primary culture. Circ Res 79:14–19

Latour I, Louw DF, Beedle AM, Hamid J, Sutherland GR, Zamponi GW (2004) Expression of T-type calcium channel splice variants in human glioma. Glia 48:112–119

Lechleiter JD, Clapham DE (1992) Molecular mechanisms of intracellular calcium excitability in X. laevis oocytes. Cell 69:283–294

Lesouhaitier O, Chiappe A, Rossier MF (2001) Aldosterone increases T-type calcium currents in human adenocarcinoma (H295R) cells by inducing channel expression. Endocrinology 142:4320–4330

Leuranguer B, Bourinet E, Lory P, Nargeot J (1998) Antisense depletion of beta-subunits fails to affect T-type calcium channels properties in a neuroblastoma cell line. Neuropharmacology 37:701–708

Lewalle JM, Cataldo D, Bajou K, Lambert CA, Foidart JM (1998) Endothelial cell intracellular Ca^{2+} concentration is increased upon breast tumor cell contact and mediates tumor cell transendothelial migration. Clin Exp Metastasis 16:21–29

Li M, Zhang M, Huang L, Zhou J, Zhuang H, Taylor JT, Keyser BM, Whitehurst RM Jr (2005) T-type Ca^{2+} channels are involved in high glucose-induced rat neonatal cardiomyocyte proliferation. Pediatr Res 57:550–556

Li W, Zhang SL, Wang N, Zhang BB, Li M (2011) Blockade of T-type Ca^{2+} channels inhibits human ovarian cancer cell proliferation. Cancer Invest 29:339–346

Lu F, Chen H, Zhou C, Liu S, Guo M, Chen P, Zhuang H, Xie D, Wu S (2008) T-type $Ca^{(2+)}$ expression in human esophageal carcinomas: a functional role in proliferation. Cell Calcium 43:49–58

Mariot P, Vanoverberghe K, Lalevee N, Rossier MF, Prevarskaya N (2002) Overexpression of an alpha 1H (Cav3.2) T-type calcium channel during neuroendocrine differentiation of human prostate cancer cells. J Biol Chem 277:10824–10833

Martini A, Bruno R, Mazzulla S, Nocita A, Martino G (2010) Angiotensin II regulates endothelial cell migration through calcium influx via T-type calcium channel in human umbilical vein endothelial cells. Acta Physiol 198:449–455

Medina RA, Owen GI (2002) Glucose transporters: expression, regulation and cancer. Biol Res 35:9–26

Mehrke G, Zong XG, Flockerzi V, Hofmann F (1994) The $Ca^{(++)}$-channel blocker Ro 40-5967 blocks differently T-type and L-type channels. J Pharmacol ExpTher 271:1483–1488

Meldrum E, Parker PJ, Carozzi A (1991) The PtdIns-PLC superfamily and signal transduction. Biochim Biophys Acta 1092:49–71

Micol V, Sanchez-Pinera P, Villalain J, de Godos A, Gomez-Fernandez JC (1999) Correlation between protein kinase C alpha activity and membrane phase behavior. Biophys J 76:916–927

Mor M, Beharier O, Levy S, Kahn J, Dror S, Blumenthal D, Gheber LA, Peretz A, Katz A, Moran A, Etzion Y (2012) ZnT-1 enhances the activity and surface expression of T-type calcium channels through activation of Ras-ERK signaling. Am J Physiol 303:C192–C203

Mulgrew CJ, Cove-Smith A, McLatchie LM, Brooks G, Shattock MJ, Hendry BM (2009) Inhibition of human mesangial cell proliferation by targeting T-type calcium channels. Nephron Exp Nephrol 113:e77–e88

Nuss HB, Houser SR (1993) T-type Ca^{2+} current is expressed in hypertrophied adult feline left ventricular myocytes. Circ Res 73:777–782

Oguri A, Tanaka T, Iida H, Meguro K, Takano H, Oonuma H, Nishimura S, Morita T, Yamasoba T, Nagai R, Nakajima T (2010) Involvement of $Ca_V3.1$ T-type calcium channels in cell proliferation in mouse preadipocytes. Am J Physiol 298:C1414–C1423

Ohkubo T, Yamazaki J (2012) T-type voltage-activated calcium channel Cav3.1, but not Cav3.2, is involved in the inhibition of proliferation and apoptosis in MCF-7 human breast cancer cells. Int J Oncol 41:267–275

Orestes P, Bojadzic D, Lee J, Leach E, Salajegheh R, Digruccio MR, Nelson MT, Todorovic SM (2011) Free radical signalling underlies inhibition of $Ca_V3.2$ T-type calcium channels by nitrous oxide in the pain pathway. J Physiol 589:135–148

Osborne CK, Boldt DH, Clark GM, Trent JM (1983) Effects of tamoxifen on human breast cancer cell cycle kinetics: accumulation of cells in early G1 phase. Cancer Res 43:3583–3585

Panner A, Cribbs LL, Zainelli GM, Origitano TC, Singh S, Wurster RD (2005) Variation of T-type calcium channel protein expression affects cell division of cultured tumor cells. Cell Calcium 37:105–119

Parkash J, Chaudhry MA, Amer AS, Christakos S, Rhoten WB (2002) Intracellular calcium ion response to glucose in beta-cells of calbindin-D28k null mutant mice and in beta HC13 cells overexpressing calbindin-D28k. Endocrine 18:221–229

Parkash J, Chaudhry MA, Rhoten WB (2004) Calbindin-D28k and calcium sensing receptor cooperate in MCF-7 human breast cancer cells. Int J Oncol 24:1111–1119

Percival RC, Yates AJ, Gray RE (1985) Mechanism of malignant hypercalcemia in carcinoma of the breast. Br Med J (Clin Res Ed) 191:776–779

Pluteanu F, Cribbs LL (2011) Regulation and function of Cav3.1 T-type calcium channels in IGF-I-stimulated pulmonary artery smooth muscle cells. Am J Physiol 300:C517–C525

Pottle JE, Sun C, Gray L, Li M (2013) Exploiting MCF-7 cells' calcium dependence with interlaced therapy. J Cancer Ther 4:32–40

Protas L, Robinson RB (2000) Mibefradil, and I(Ca, T) blocker effectively blocks I(Ca, L) in rabbit sinus node cells. Eur J Pharmacol 401:27–30

Rasmussen CD, Means AR (1989) Calmodulin is required for cell-cycle progression during G1 and mitosis. EMBO J 8:73–82

Reddy GP, Reed WC, Sheehan E, Sacks DB (1992) Calmodulin-specific monoclonal antibodies inhibit DNA replication in mammalian cells. Biochemistry 31:10426–10430

Richard S, Neveu D, Carnac G, Bodin P, Travo P, Nargeot J (1992) Differential expression of voltage-gated $Ca^{(2+)}$-currents in cultivated aortic myocytes. Biochim Biophys Acta 1160:95–104

Rodman DM, Reese K, Harral J, Fouty B, Wu S, West J, Hoebt-Miller M, Tada Y, Li KX, Cool C, Fagan K, Cribbs L (2005) Low-voltage-activated (T-type) calcium channels control proliferation of human pulmonary artery myocytes. Circ Res 96:864–872

Rogers S, Macheda ML, Docherty SE, Carty MD, Henderson MA, Soeller WC, Gibbs EM, James DE, Best JD (2002) Identification of a novel glucose transporter-like protein—GLUT-12. Am J Physiol 282:E733–E738

Sanders JL, Chattopadhyay N, Kifor O, Yamaguchi T, Butters RR, Brown EM (2000) Extracellular calcium-sensing receptor expression and its potential role in regulating parathyroid hormone-related peptide secretion in human breast cancer cell lines. Endocrinology 141:4357–4364

Santoni G, Santoni M, Nabissi M (2012) Functional role of T-type calcium channels in tumour growth and progression: prospective in cancer therapy. Br J Pharmacol 166:1244–1246

Schmitt R, Clozel JP, Iberg N, Buhler FR (1995) Mibefradil prevents neointima formation after vascular injury in rats. Possible role of the blockade of the T-type voltage-operated calcium channel. Arterioscler Thromb Vasc Biol 8:1161–1165

Shimoyama Y, Hirohashi S (1991) Cadherin intercellular adhesion molecule in hepatocellular carcinomas: loss of E-cadherin expression in an undifferentiated carcinoma. Cancer Lett 57:131–135

Siitonen SM, Isola JJ, Rantala IS, Helin HJ (1993) Intratumor variation in cell proliferation in breast carcinoma as determined by antiproliferating cell nuclear antigen monoclonal antibody and automated image analysis. Am J Clin Pathol 99:226–231

Silver IA, Murrils RJ, Etherington DJ (1998) Microelectrode studies on the acid microenvironment beneath adherent macrophages and osteroclasts. Exp Cell Res 175:266–276

Smith TA (1999) Facilitative glucose transporter expression in human cancer tissue. Br J Biomed Sci 56:285–292

Steinhardt RA, Alderton J (1988) Intracellular free calcium rise triggers nuclear envelope breakdown in the sea urchin embryo. Nature 332:364–366

Stuart CA, Howell ME, Zhang Y, Yin D (2009) Insulin-stimulated translocation of glucose transporter (GLUT) 12 parallels that of GLUT4 in normal muscle. J Clin Endocrinol Metab 94:3535–3542

Taylor JT, Huang L, Pottle JE, Liu K, Yang Y, Zeng X, Keyser BM, Agrawal KC, Hansen JB, Li M (2008a) Selective blockade of T-type Ca^{2+} channels suppresses human breast cancer cell proliferation. Cancer Lett 267:116–124

Taylor JT, Zeng XB, Pottle JE, Lee K, Wang AR, Yi SG, Scruggs JAS, Sikka SS, Li M (2008b) Calcium signaling and T-type calcium channels in cancer cell cycling. World J Gastroenterol 14:4984–4991

Toyota M, Ho C, Ohe-Toyota M, Baylin SB, Issa JP (1999) Inactivation of CACNA1G, a T-type calcium channel gene, by aberrant methylation of its 5' CpG island in human tumors. Cancer Res 59:4535–4541
Trump BF, Berezesky IK (1995) Calcium-mediated cell injury and cell death. FASEB J 9:219–228
Valerie NC, Dziegielewska B, Hosing AS, Augustin E, Gray LS, Brautigan DL, Larner JM, Dziegielewski J (2013) Inhibition of T-type calcium channels disrupts Akt signaling and promotes apoptosis in glioblastoma cells. Biochem Pharmacol 85:888–897
Villame J, Massicotte J, Jasmin G, Dumont L (2001) Effects of mibefradil, a T- and L-type calcium channel blocker, on cardiac remodeling in the UM-X7.1 cardiomyopathic hamster. Cardiovasc Drugs Ther 15:41–48
Volm M, Wayss K (1975) Failure to partial synchronization with hydroxyurea to improve results of chemotherapy of solid neurosarcoma in the rat. Dtsch Med Wochenschr 100:2102–2104
Wang XT, Nagaba Y, Cross HS, Wrba F, Zhang L, Guggino SE (2000) The mRNA of L-type calcium channel elevated in colon cancer. Am J Pathol 157:1549–1562
Wang YQ, Brooks G, Zhu CB, Yuan WZ, Li YQ, Wu XS (2002) Functional analysis of the human T-type calcium channel alpha 1H subunit gene in cellular proliferation. Yichuan Xuebao 29:659–665
Whitfield JF (1992) Calcium signals and cancer. Crit Rev Oncog 3:55–90
Wildburger NC, Lin-Ye A, Baird MA, Lei D, Bao J (2009) Neuroprotective effects of blockers for T-type calcium channels. Mol Neurodegener 4:44
Wolfe JT, Wang H, Howard J, Garrison JC, Barrett PQ (2003) T-type calcium channel regulation by specific G-protein betagamma subunits. Nature 424:209–213
Wu S, Zhang M, Vest PA, Bhattacharjee A, Liu L, Li M (2000) A mibefradil metabolite is a potent intracellular blocker of L-type Ca^{2+} currents in pancreatic β-Cells. J Pharmacol Exp Ther 292:939–943
Wyatt CN, Page KM, Berrow NS, Brice NL, Dolphin AC (1998) The effect of overexpression of auxiliary Ca2+ channel subunits on native Ca2+ channel currents in undifferentiated mammalian NG108-15 cells. J Physiol 510:347–360
Yin Y, Henzl MT, Lorber B, Nakazawa T, Thomas TT, Jiang F, Langer R, Benowitz LI (2006) Oncomodulin is a macrophage-derived signal for axon regeneration in retinal ganglion cells. Nat Neurosci 9:843–852
Zamponi GW, Snutch TP (1998) Modulation of voltage-dependent calcium channels by G proteins. Neurobiology 8:351–356
Zawadzki A, Liu Q, Wang Y, Melander A, Jeppsson B, Thorlacius H (2008) Verapamil inhibits L-type channel mediated apoptosis in human colon cancer cells. Dis Colon Rectum 51:1696–1702
Zeitler H, Ko Y, Glodny B, Totzke G, Appenheimer M, Sachinidis A, Vetter H (1997) Cell cycle arrest in G0/G1 phase of growth-factor induced endothelial cell proliferation by various calcium channel blockers. Cancer Detect Prev 21:332–339
Zhang Y, Zhang J, Jiang D, Zhang D, Qian Z, Liu C, Tao J (2012) Inhibition of T-type Ca^{2+} channels by endostatin attenuates human glioblastoma cell proliferation and migration. Br J Pharmacol 166:1247–1260
Zhuang H, Bhattacharjee A, Hu F, Zhang M, Goswami T, Wang L, Wu S, Berggren PO, Li M (2000) Cloning of a T-type Ca^{2+} channel isoform in insulin secreting cells. Diabetes 49:59–64
Zucker RS, Steinhardt RA (1978) Prevention of the cortical reaction in fertilized sea urchin eggs by injection of calcium-chelating ligands. Biochim Biophys Acta 541:459–466

T-Type Calcium Channels and Management of Pain

9

Brian M. Keyser

Abstract

T-type Ca^{2+} channels exhibit small, transient currents that are relatively insensitive to dihydropyridines and exhibit "window current" in the physiological range for voltage. A role for T-type Ca^{2+} channels in nociceptor sensitization was reported in 2001, an effect confirmed in studies using $Ca_v3.1$ and $Ca_v3.2$ knockout animals. The $Ca_v3.2$ channel is involved in both central and peripheral termini of primary afferent neurons and is now considered an important player in the processing of pain. The analgesic effect of T-type calcium channels has been tested in different pain models. In addition to neuropathic pain reduction, the inhibition of T-type calcium channels is efficacious in formalin injection into the hind paw, which models an acute inflammatory pain. T-type Ca^{2+} channel blockers also produce a dose-dependent reduction in IBS-induced pain in a rat model of irritable bowel syndrome (IBS). T-type Ca^{2+} channel antagonists present obvious advantages over opioid analgesics; they directly act on the cytoplasmic membrane of neurons and most likely from the extracellular side. Therefore their antinociceptive effect may not cause profound drug tolerance and dependence.

9.1 Introduction

Voltage-activated calcium channels are a family of calcium channels that are activated by changes in the electrical potential of the cell membrane and are composed of six transmembrane spanning domains, with the fourth transmembrane domain acting as the voltage sensor. These channels are found in a multitude of

B.M. Keyser (✉)
Cellular and Molecular Biology Branch, U.S. Army Medical Research Institute of Chemical Defense Research Division, 3100 Ricketts Point Road, Aberdeen Proving Ground, MD 21015, USA

S.W. Schaffer, M. Li (eds.), *T-type Calcium Channels in Basic and Clinical Science*,
DOI 10.1007/978-3-7091-1413-1_9

cells, including the neurons, heart, brain, muscle, and pancreatic β-cells, where they allow calcium entry into the cell. Although these channels only remain activate for a short duration following depolarization of the cell, they play a key role in regulating $[Ca^{2+}]_i$.

The nomenclature of these calcium channels has been reclassified based on their amino acid sequences and the homology with other calcium channels. The major families of calcium channels are designated Ca_v1, Ca_v2, and Ca_v3. The members of Ca_v1 are as follows (with the old nomenclature in parenthesis): $Ca_v1.1$ (α_1S), $Ca_v1.2$ (α_1C), $Ca_v1.3$ (α_1D), and $Ca_v1.4$ (α_1F). The other two families contain three members, each designated as $Ca_v2.1$ (α_1A), $Ca_v2.2$ (α_1B), $Ca_v2.3$ (α_1E), $Ca_v3.1$ (α_1G), $Ca_v3.2$ (α_1H), and $Ca_v3.3$ (α_1I). All the channels in each family share common elements, such as all the channels in the Ca_v1 family are important in the excitation-contraction coupling in the skeletal, cardiac, and smooth muscle; the Ca_v2 family can all be antagonized by various toxins, whereas the Ca_v3 family contains all of the low-voltage-activated calcium channels.

The Ca_v3 family of voltage-activated calcium channels contains all of the low-voltage-activated calcium channels. These channels can be found in the heart, kidney, nervous tissue, brain, smooth muscle, sperm, and many endocrine tissues. All of the low-voltage-activated calcium channels have common features that set them apart from the other high-voltage-activated calcium channels (Ca_v1 and 2). They are characterized by small, transient currents, relative insensitivity to dihydropyridines and the presence of a "window current" at the physiological range for voltages. Window current is a phenomenon in which the activation and inactivation curves overlap and a small number of channels remain in a persistently open state. This window current has been shown to be a mechanism for calcium entry into the cell, which triggers hormone release by the heart, kidney, and neuroendocrine tissues (Yang and Berggren 2006; Belardetti and Zamponi 2008; Marcantoni et al. 2008). T-type calcium channels have been proposed to play a role in pathophysiological processes, including pain (Iftinca 2011) and diabetic neuropathy (Calcutt 2013).

9.2 T-Type Calcium Channel in Nociception

The involvement of T-type Ca^{2+} channels in nociceptor sensitization was first reported in 2001 (Todorovic et al. 2001) and has been confirmed in studies using $Ca_v3.1$and $Ca_v3.2$ knockout animals (Shin et al. 2008). The $Ca_v3.2$ channel, which is involved in both central and peripheral termini of primary afferent neurons (Maeda et al. 2009; Jacus et al. 2012; Weiss et al. 2012), is now thought to play an important role in the processing of pain (Sekiguchi and Kawabata 2013).

The regulation of $Ca_v3.2$ channels has been thoroughly investigated in excitable cells for the last 10 years (Huc et al. 2009). Many of the intracellular second messenger pathways exert their regulatory effects on the intracellular loop between domains II and III (loops II–III). The messengers of these pathways include Gβγ,

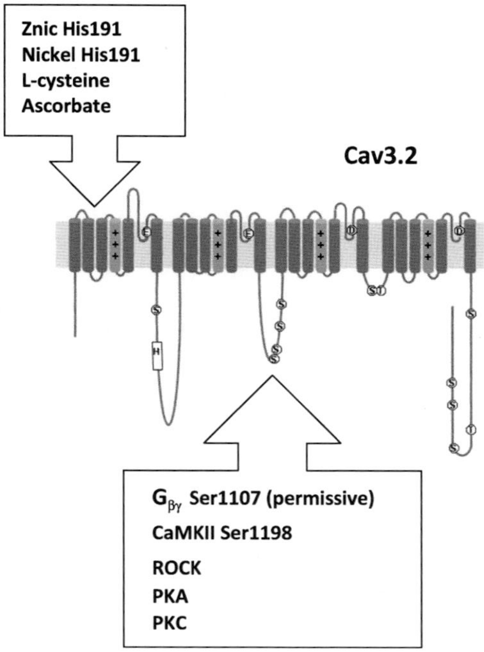

Fig. 9.1 The scheme shows the major intracellular and extracellular regulatory sites of $Ca_v3.2$ T-type Ca^{2+} channels

GTPase, PKA, PKC, and CaMKII; the extracellular signals exert their effect directly on loops III–IV of domain I as shown in Fig. 9.1.

In addition to roles in processing pain in the periphery, $Ca_v3.2$ activation in the CNS (paraventricular thalamus) has been shown to contribute to acid-induced chronic hyperalgesia (Chen et al. 2010). The $Ca_v3.2$ T-type Ca^{2+} channel is also involved in nociception and tactile reticular thalamic neuronal burst firing (Liao et al. 2011).

The inhibition of these channels has provided more evidence for the important role of T-type calcium channels in nociception. Using antisense oligodeoxynucleotides against the whole T-type calcium channel family, one minimizes noxious stimuli-mediated nociception in healthy and mononeuropathic rats (Bourinet et al. 2005). Moreover, the T-type calcium channel antagonist, mibefradil, was found to provide peripheral thermal analgesia and mechanical nociception in rats (Todorovic et al. 2001, 2002). Another T-type calcium channel antagonist, ethosuximide, was able to produce similar results in thermal and mechanical models of pain (Dogrul et al. 2003; Flatters and Bennett 2004).

9.3 T-Type Calcium Channels in Pain Models

T-type calcium channels appear to play a more important role in neuropathic pain than nociception. The expression of these channels and current density are upregulated in models of chronic constrictive injury (CCI) (Jagodic et al. 2008;

Takahashi et al. 2010). When these channels are inhibited both pharmacologically and at the molecular level, there is a significant reduction in pain (Flatters and Bennett 2004; Sekiguchi and Kawabata 2013; Jarvis et al. 2014). These papers mainly examine the role of $Ca_v3.2$ in neuropathic pain; however, both $Ca_v3.1$ and $Ca_v3.3$ have been implicated in this process. $Ca_v3.1$-defective mice attenuate pain after L5 spinal nerve ligation, whereas the mRNA of $Ca_v3.3$ is upregulated in a rat model of chronic compression of dorsal root ganglion (DRG) neurons (Na et al. 2008; Wen et al. 2010; Yue et al. 2013). The antagonism of these channels with either mibefradil or 3,5-dichloro-*N*-[1-(2,2-dimethyl-tetrahydro-pyran-4-ylmethyl)-4-fluoro-piperidin-4-ylmethyl]-benzamide (TTA-P2) is also able to alleviate the pain (Yue et al. 2013; Choe et al. 2011).

In addition to neuropathic pain reduction, inhibition of T-type calcium channels is also efficacious in formalin injection into the hind paw, which models an acute inflammatory pain. After formalin is injected two phases are observed: the first phase is short (~10 min), followed by a brief (~5 min) period of quiescence, after which the second phase of hind limb shaking and licking lasts for an additional 50 min (National Research Council 2009). The response to both of these phases is reduced upon treatment with new T-type calcium channel antagonists and in experiments using $Ca_v3.2$ knockout mice (Choi et al. 2007; Kam et al. 2010; Lee et al. 2010; Choe et al. 2011).

Irritable bowel syndrome (IBS) is a gastrointestinal (GI) disorder in which recurrent digestive symptoms develop without overt pathology. IBS causes significant abdominal pain although presently there is no treatment for this pain. Recent evidence has shown that $Ca_v3.2$ is involved in the pain associated with IBS in vivo (Marger et al. 2011). The use of TTA-A2 (structurally similar to TTA-P2) in a rat model of IBS was shown to produce a dose-dependent reduction in IBS-induced pain (Francois et al. 2013). These preliminary results show the possible use of T-type calcium channel inhibitors in reducing IBS-induced pain.

9.4 Dual Use of T-Type Calcium Channels in Diseases

In medicine today, it is desirable to use one drug which can affect multiple symptoms of a disorder, a so-called magic bullet. The discovery of drugs that serve as magic bullets promotes patient compliance and therefore increases treatment success (Atreja et al. 2005). Since T-type calcium channels are involved in the pathogenesis of multiple diseases (i.e., cancer and type 2 diabetes mellitus) and in the pain associated with those diseases, antagonists of these channels may offer a unique target for achieving the desired outcome.

The $Ca_v3.2$ channels appear to play an important role in the pain associated with diabetic neuropathy (Jagodic et al. 2007; Messinger et al. 2009; Calcutt 2013; Khomula et al. 2013). The use of either mibefradil at clinically relevant doses or the appropriate antisense oligonucleotide alleviates heat, cold, and mechanical hypersensitivity in streptozotocin (STZ)-injected rats (Messinger et al. 2009; Obradovic et al. 2014). These results have been replicated in another model of type 2 diabetes mellitus, the leptin-deficient (ob/ob) mouse, showing the importance of T-type calcium channels in diabetic neuropathy (Latham et al. 2009). These channels are also involved in type 2 diabetes mellitus, as the antagonism of these channels both in vitro and in vivo improves insulin secretion (Keyser et al. 2014). Therefore, the use of an antagonist of T-type calcium channels could treat both diabetic neuropathic pain and the symptoms of type 2 diabetes mellitus.

The incidence of pain with the use of anticancer drugs is 64 % and is detected during all phases of cancer (early and metastatic) (Ripamonti et al. 2011). The management of this pain, for which there is presently no treatment, could provide relief to these patients and increase their well-being. The use of various T-type calcium channel antagonists has shown to reduce the pain induced by paclitaxel or vincristine (Okubo et al. 2011; Flatters and Bennett 2004). These channels are upregulated during the fast replication phase of cancer cells; antagonism of these channels inhibits cell proliferation (Taylor et al. 2008a, b; Li et al. 2011; Ohkubo and Yamazaki 2012; Gray et al. 2013). The inhibition of this channel in vivo showed reduction in transplanted tumor size in mice (Pottle et al. 2013). In addition to being effective when applied as the sole treatment, antagonists also enhance the anticancer effects of conventional radio- and chemotherapy (Pottle et al. 2013; Dziegielewska et al. 2014). These results suggest that the use of T-type calcium channel antagonists in the management of pain also exhibits beneficial effects against the disease itself.

Conclusions

Inadequate treatment of pain is widespread throughout surgical wards, intensive care units, accident and emergency departments, in general practice, and in the management of all forms of chronic pain, including cancer pain and end of life care (Rupp and Delaney 2004; Smith and Toonen 2007). Current management of chronic and severe pain mainly relies on the use of opioid analgesics, which produce a wide spectrum of unwanted side effects, including respiratory depression, nausea, vomiting, dizziness, mental clouding, dysphoria, pruritus, constipation, increased pressure in the biliary tract, urinary retention, and hypotension. Furthermore, opioids tend to induce drug dependence and tolerance, effects that involve receptor downregulation and second messenger compensation (Bie et al. 2005; Pradhan et al. 2010).

In contrast, T-type Ca^{2+} channel antagonists present obvious advantages over opioids; they directly act on the cytoplasmic membrane of neurons, presumably from the extracellular side. Therefore, their antinociceptive effect is unlikely to cause profound drug tolerance and dependence. One of the nonselective T-type Ca^{2+} channel blockers, mibefradil, was once on the market for treatment of hypertension, and clinical reports found that it was well tolerated (Brogden and Markham 1997). The most adverse events were either related to the vasodilatory action of the drug (Brogden and Markham 1997). Therefore T-type Ca^{2+} channels, specifically the $Ca_v3.2$ isoform, are promising targets for future treatment of pain. More attention should focus on the development of more selective and efficacious T-type Ca^{2+} channel antagonists.

References

Atreja A, Bellam N, Levy SR (2005) Strategies to enhance patient adherence: making it simple. Med Gen Med 7:4

Belardetti F, Zamponi GW (2008) Linking calcium-channel isoforms to potential therapies. Curr Opin Investig Drugs 9:707–715

Bie B, Peng Y, Zhang Y, Pan ZZ (2005) cAMP-mediated mechanisms for pain sensitization during opioid withdrawal. J Neurosci 25:3824–3832

Bourinet E, Alloui A, Monteil A, Barrere C, Couette B, Poirot O, Pages A, McRory J, Snutch TP, Eschalier A, Nargeot J (2005) Silencing of the Cav3.2 T-type calcium channel gene in sensory neurons demonstrates its major role in nociception. EMBO J 24:315–324

Brogden RN, Markham A (1997) Mibefradil—a review of its pharmacodynamic and pharmacokinetic properties, and therapeutic efficacy in the management of hypertension and angina pectoris. Drugs 54:774–793

Calcutt NA (2013) Location, location, location?: Is the pain of diabetic neuropathy generated by hyperactive sensory neurons? Diabetes 62:3658–3660

Chen WK, Liu IY, Chang YT, Chen YC, Chen CC, Yen CT, Shin HS, Chen CC (2010) Ca(v)3.2 T-type Ca^{2+} channel-dependent activation of ERK in paraventricular thalamus modulates acid-induced chronic muscle pain. J Neurosci 30:10360–10368

Choe W, Messinger RB, Leach E, Eckle VS, Obradovic A, Salajegheh R, Jevtovic-Todorovic V, Todorovic SM (2011) TTA-P2 is a potent and selective blocker of T-type calcium channels in rat sensory neurons and a novel antinociceptive agent. Mol Pharmacol 80:900–910

Choi S, Na HS, Kim J, Lee J, Lee S, Kim D, Park J, Chen CC, Campbell KP, Shin HS (2007) Attenuated pain responses in mice lacking Ca(V)3.2 T-type channels. Genes Brain Behav 6:425–431

Dogrul A, Gardell LR, Ossipov MH, Tulunay FC, Lai J, Porreca F (2003) Reversal of experimental neuropathic pain by T-type calcium channel blockers. Pain 105:159–168

Dziegielewska B, Gray LS, Dziegielewski J (2014) T-type calcium channels blockers as new tools in cancer therapies. Pflugers Arch 466:801–810

Flatters SJ, Bennett GJ (2004) Ethosuximide reverses paclitaxel- and vincristine-induced painful peripheral neuropathy. Pain 109:150–161

Francois A, Kerckhove N, Meleine M, Alloui A, Barrere C, Gelot A, Uebele VN, Renger JJ, Eschalier A, Ardid D, Bourinet E (2013) State-dependent properties of a new T-type calcium channel blocker enhance Ca(V)3.2 selectivity and support analgesic effects. Pain 154:283–293

Gray LS, Schiff D, Macdonald TL (2013) A model for the regulation of T-type Ca(2+) channels in proliferation: roles in stem cells and cancer. Expert Rev Anticancer Ther 13:589–595

Huc S, Monteil A, Bidaud I, Barbara G, Chemin J, Lory P (2009) Regulation of T-type calcium channels: signaling pathways and functional implications. Biochem Biophys Acta 1793:947–952

Iftinca MC (2011) Neuronal T-type calcium channels: what's new? Iftinca: T-type channel regulation. J Med Life 4:126–138

Jacus MO, Uebele VN, Renger JJ, Todorovic SM (2012) Presynaptic Cav3.2 channels regulate excitatory neurotransmission in nociceptive dorsal horn neurons. J Neurosci 32:9374–9382

Jagodic MM, Pathirathna S, Nelson MT, Mancuso S, Joksovic PM, Rosenberg ER, Bayliss DA, Jevtovic-Todorovic V, Todorovic SM (2007) Cell-specific alterations of T-type calcium current in painful diabetic neuropathy enhance excitability of sensory neurons. J Neurosci 27:3305–3316

Jagodic MM, Pathirathna S, Joksovic PM, Lee W, Nelson MT, Naik AK, Su P, Jevtovic-Todorovic V, Todorovic SM (2008) Upregulation of the T-type calcium current in small rat sensory neurons after chronic constrictive injury of the sciatic nerve. J Neurophysiol 99:3151–3156

Jarvis MF, Scott VE, McGaraughty S, Chu KL, Xu J, Niforatos W, Milicic I, Joshi S, Zhang Q, Xia Z (2014) A peripherally acting, selective T-type calcium channel blocker, ABT-639, effectively reduces nociceptive and neuropathic pain in rats. Biochem Pharmacol 89:536–544

Kam YL, Rhee HK, Rhim H, Back SK, Na HS, Choo HY (2010) Synthesis and T-type calcium channel blocking activity of novel diphenylpiperazine compounds, and evaluation of in vivo analgesic activity. Bioorg Med Chem 18:5938–5944

Keyser BM, Taylor JT, Choi S-K, Lu Y, Bhattacharjee A, Huang L, Pottle J, Matrougui K, Xu Z, Li M (2014) Role of T-type Ca^{2+} channels in basal $[Ca^{2+}]_i$ regulation and basal insulin secretion in rat islet cells. Curr Trend Endocrinol

Khomula EV, Viatchenko-Karpinski VY, Borisyuk AL, Duzhyy DE, Belan PV, Voitenko NV (2013) Specific functioning of Cav3.2 T-type calcium and TRPV1 channels under different types of STZ-diabetic neuropathy. Biochim Biophys Acta 1832:636–649

Latham JR, Pathirathna S, Jagodic MM, Choe WJ, Levin ME, Nelson MT, Lee WY, Krishnan K, Covey DF, Todorovic SM, Jevtovic-Todorovic V (2009) Selective T-type calcium channel blockade alleviates hyperalgesia in ob/ob mice. Diabetes 58:2656–2665

Lee MJ, Shin TJ, Lee JE, Choo H, Koh HY, Chung HJ, Pae AN, Lee SC, Kim HJ (2010) KST5468, a new T-type calcium channel antagonist, has an antinociceptive effect on inflammatory and neuropathic pain models. Pharmacol Biochem Behav 97:198–204

Li W, Zhang SL, Wang N, Zhang BB, Li M (2011) Blockade of T-type Ca^{2+} channels inhibits human ovarian cancer cell proliferation. Cancer Invest 29:339–346

Liao YF, Tsai ML, Chen CC, Yen CT (2011) Involvement of the Cav3.2 T-type calcium channel in thalamic neuron discharge patterns. Mol Pain 7:43

Maeda Y, Aoki Y, Sekiguchi F, Matsunami M, Takahashi T, Nishikawa H, Kawabata A (2009) Hyperalgesia induced by spinal and peripheral hydrogen sulfide: evidence for involvement of Cav3.2 T-type calcium channels. Pain 142:127–132

Marcantoni A, Carabelli V, Comunanza V, Hoddah H, Carbone E (2008) Calcium channels in chromaffin cells: focus on L and T types. Acta Physiol 192:233–246

Marger F, Gelot A, Alloui A, Matricon J, Ferrer JF, Barrere C, Pizzoccaro A, Muller E, Nargeot J, Snutch TP, Eschalier A, Bourinet E, Ardid D (2011) T-type calcium channels contribute to colonic hypersensitivity in a rat model of irritable bowel syndrome. Proc Natl Acad Sci USA 108:11268–11273

Messinger RB, Naik AK, Jagodic MM, Nelson MT, Lee WY, Choe WJ, Orestes P, Latham JR, Todorovic SM, Jevtovic-Todorovic V (2009) In vivo silencing of the Ca(V)3.2 T-type calcium channels in sensory neurons alleviates hyperalgesia in rats with streptozocin-induced diabetic neuropathy. Pain 145:184–195

Na HS, Choi S, Kim J, Park J, Shin HS (2008) Attenuated neuropathic pain in Cav3.1 null mice. Mol Cells 25:242–246

National Research Council (2009) Recognition and alleviation of pain in laboratory animals. The National Academies Press, Washington, DC

Obradovic A, Hwang SM, Scarpa J, Hong SJ, Todorovic SM, Jevtovic-Todorovic V (2014) CaV3.2 T-type calcium channels in peripheral sensory neurons are important for mibefradil-induced reversal of hyperalgesia and allodynia in rats with painful diabetic neuropathy. PLoS One 9:e91467

Ohkubo T, Yamazaki J (2012) T-type voltage-activated calcium channel Cav3.1, but not Cav3.2, is involved in the inhibition of proliferation and apoptosis in MCF-7 human breast cancer cells. Int J Oncol 41:267–275

Okubo K, Takahashi T, Sekiguchi F, Kanaoka D, Matsunami M, Ohkubo T, Yamazaki J, Fukushima N, Yoshida S, Kawabata A (2011) Inhibition of T-type calcium channels and hydrogen sulfide-forming enzyme reverses paclitaxel-evoked neuropathic hyperalgesia in rats. Neuroscience 188:148–156

Pottle J, Sun C, Gray L, Li M (2013) Exploiting MCF-7 cells' calcium dependence with interlaced therapy. J Cancer Ther 4:32–40

Pradhan AA, Walwyn W, Nozaki C, Filliol D, Erbs E, Matifas A, Evans C, Kieffer BL (2010) Ligand-directed trafficking of the δ-opioid receptor in vivo: two paths toward analgesic tolerance. J Neurosci 30:16459–16468
Ripamonti CI, Bandieri E, Roila F, Group EGW (2011) Management of cancer pain: ESMO Clinical Practice Guidelines. Ann Oncol 22(Suppl 6):vi69–77
Rupp T, Delaney KA (2004) Inadequate analgesia in emergency medicine. Ann Emerg Med 43:494–503
Sekiguchi F, Kawabata A (2013) T-type calcium channels: functional regulation and implication in pain signaling. J Pharmacol Sci 122:244–250
Shin HS, Cheong EJ, Choi S, Lee J, Na HS (2008) T-type Ca^{2+} channels as therapeutic targets in the nervous system. Curr Opin Pharmacol 8:33–41
Smith GF, Toonen TR (2007) Primary care of the patient with cancer. Am Fam Physician 75:1207–1214
Takahashi T, Aoki Y, Okubo K, Maeda Y, Sekiguchi F, Mitani K, Nishikawa H, Kawabata A (2010) Upregulation of Ca(v)3.2 T-type calcium channels targeted by endogenous hydrogen sulfide contributes to maintenance of neuropathic pain. Pain 150:183–191
Taylor JT, Huang L, Pottle JE, Liu K, Yang Y, Zeng X, Keyser BM, Agrawal KC, Hansen JB, Li M (2008a) Selective blockade of T-type Ca^{2+} channels suppresses human breast cancer cell proliferation. Cancer Lett 267:116–124
Taylor JT, Zeng XB, Pottle JE, Lee K, Wang AR, Yi SG, Scruggs JA, Sikka SS, Li M (2008b) Calcium signaling and T-type calcium channels in cancer cell cycling. World J Gastroenterol 14:4984–4991
Todorovic SM, Jevtovic-Todorovic V, Meyenburg A, Mennerick S, Perez-Reyes E, Romano C, Olney JW, Zorumski CF (2001) Redox modulation of T-type calcium channels in rat peripheral nociceptors. Neuron 31:75–85
Todorovic SM, Meyenburg A, Jevtovic-Todorovic V (2002) Mechanical and thermal antinociception in rats following systemic administration of mibefradil, a T-type calcium channel blocker. Brain Res 951:336–340
Weiss N, Hameed S, Fernandez-Fernandez JM, Fablet K, Karmazinova M, Poillot C, Proft J, Chen L, Bidaud I, Monteil A, Huc-Brandt S, Lacinova L, Lory P, Zamponi GW, De Waard M (2012) A Ca(v)3.2/syntaxin-1A signaling complex controls T-type channel activity and low-threshold exocytosis. J Biol Chem 287:2810–2818
Wen XJ, Xu SY, Chen ZX, Yang CX, Liang H, Li H (2010) The roles of T-type calcium channel in the development of neuropathic pain following chronic compression of rat dorsal root ganglia. Pharmacology 85:295–300
Yang SN, Berggren PO (2006) The role of voltage-gated calcium channels in pancreatic beta-cell physiology and pathophysiology. Endocr Rev 27:621–676
Yue J, Liu L, Liu Z, Shu B, Zhang Y (2013) Upregulation of T-type Ca^{2+} channels in primary sensory neurons in spinal nerve injury. Spine 38:463–470

Role of T-Type Ca^{2+} Channels in Basal Insulin Release

10

Ming Li

Abstract

Increases in plasma glucose concentration enhance the expression and activity of T-type Ca^{2+} channels in insulin-secreting pancreatic β-cells. The resulting elevation in T-type Ca^{2+} channel activity results in an elevation in the basal $[Ca^{2+}]_i$ of β-cells. High $[Ca^{2+}]_i$ in turn causes the internalization of L-type Ca^{2+} channels from cytoplasmic membrane, which contributes to glucose toxicity in β-cells. Increased activity of T-type Ca^{2+} channels, along with elevated $[Ca^{2+}]_i$, promotes low threshold exocytosis of insulin under non-stimulus conditions. Emptying immediately releasable secretory granules and internalization of L-type Ca^{2+} channels result in defective first phase of glucose-induced insulin secretion from the β-cells. Consequently, this defect in glucose-induced insulin secretion augments the plasma glucose concentration and slows glucose uptake by peripheral tissues. The overall effect of the glucose-T-type Ca^{2+} channel interaction is enhanced basal insulin levels but decreased glucose-induced insulin secretion. Thus, the T-type Ca^{2+} channel overexpression mechanism could play a role in the development of hyperinsulinemia and insulin resistance, which occurs during the early phase of type 2 diabetes mellitus.

10.1 Introduction

Opposed to high voltage-activated Ca^{2+} channels, the expression of T-type Ca^{2+} channels in pancreatic β-cells is upregulated by high concentrations of glucose. What are the physiological consequences of upregulated T-type Ca^{2+} channels in β-cells? Do these changes result in elevations in intracellular calcium ($[Ca^{2+}]_i$) and

M. Li (✉)
Department of Physiology, Tulane University Health Sciences Center, New Orleans, LA 70112, USA
e-mail: mli@tulane.edu

S.W. Schaffer, M. Li (eds.), *T-type Calcium Channels in Basic and Clinical Science*,
DOI 10.1007/978-3-7091-1413-1_10

in spontaneous electrical activity of the cells? Do these changes have an impact on insulin release by β-cells, even under non-stimulatory conditions? Could the change in basal insulin release play a role in the development of insulin resistance? Finally, could T-type Ca^{2+} channels represent new targets for treatment of type 2 diabetes mellitus?

Recent research has yielded interesting results which may help answer these questions. The data support the hypothesis that T-type Ca^{2+} channels are involved in the process of insulin release under non-stimulatory conditions.

10.2 T-Type Ca^{2+} Channels Are Upregulated by Chronic High Glucose

Misler et al. (1992b) conclusively demonstrated that human pancreatic β-cells express T-type Ca^{2+} channels. In their experiment, they recorded a large voltage-dependent Ca^{2+} current that peaked at −35 mV and became inactive at −50 mV. This current could be distinguished from high voltage-activated Ca^{2+} current by its insensibility to Bay K8644. Prior to the classical 1992 study, similar types of current were reported using mice (Hopkins et al. 1991) and rat (Ashcroft et al. 1990; Sala and Matteson 1990) pancreatic β-cells. Although the existence of T-type Ca^{2+} currents was commonly accepted, the function of the current in insulin release had generated a conundrum, especially in comparison with the well-established pivotal function of L-type Ca^{2+} channels in glucose-stimulated insulin secretion (Misler et al. 1992a). The difficulty came partially from the fact that the appearance and size of this type of Ca^{2+} current had been inconsistent, depending on cell types and experimental conditions. Indeed, in many whole-cell recordings, this current was either minimal or absent. How can a channel have an important biological function if it is not regularly expressed?

In order to answer this question, we need to ask another question, which is: How is the expression of T-type Ca^{2+} channels regulated by high concentrations of glucose? To date there have been three T-type Ca^{2+} channel genes identified: α_1G (Ca_v3.1) (Perez-Reyes et al. 1998), α_1H (Ca_v3.2) (Cribbs et al. 1998), and α_1I (Ca_v3.3) (Lee et al. 1999). α_1G has been re-cloned from rat insulin secretion cells (Zhuang et al. 2000), and the mRNA of α_1G and α_1H has been detected in rat islets (Yang and Berggren 2005; Keyser et al. 2014). In contrast to the downregulation of L-type Ca^{2+} channel expression and activity in diabetic rat pancreatic β-cells (Iwashima et al. 1993; Roe et al. 1996), T-type Ca^{2+} current density and mRNA (α_1G and α_1H) levels increased markedly after exposure to 11.1 mM glucose for 48 h (Zhang et al. 2000; Keyser et al. 2014). These results are consistent with T-type Ca^{2+} channel regulation in neonatal cardiomyocytes, in which the T-type Ca^{2+} current density also increased significantly after chronic high glucose treatment (Li et al. 2005). Therefore, chronic high glucose increases both messenger expression and electrical current of T-type Ca^{2+} channels.

T-type Ca^{2+} channels assume a more important physiological and/or pathological role in pancreatic β-cells when plasma glucose levels are elevated. More specifically, the results suggest that under hyperglycemic conditions, the threshold of the action potential spikes decreases and the frequency of the spikes increases in these cells. We also suggest that the basal level of $[Ca^{2+}]_i$ increases in these cells.

10.3 T-Type Ca^{2+} Currents Enhance Excitability of Pancreatic β-Cells

A double-pulse protocol of current-clamp recording in a rat insulin-secreting cell line, INS-1, was employed to determine the contribution of T-type Ca^{2+} channels in the modulation of electrical activity and in stimulus-secretion coupling (Bhattacharjee et al. 1997). In this protocol, the electrical activity generated by two separated positive current pulses was compared with or without a negative current injection immediately before the second pulse (Fig. 10.1). The negative current hyperpolarized the membrane and lead to the recovery of inactive T-type Ca^{2+} channels. The results showed that activation of T-type Ca^{2+} channels lowers the threshold of the action potential (represented by the decreased latency of the first spike, Fig. 10.1b). When the threshold was lowered, the frequency of the action potentials increased. This result was abolished by administration of nickel chloride, T-type Ca^{2+} channel blocker. Moreover, application of a high-frequency stimulus caused a greater change in membrane capacitance (ΔCm) than the application of a low-frequency stimulus, suggesting a higher exocytotic secretion. Nickel chloride

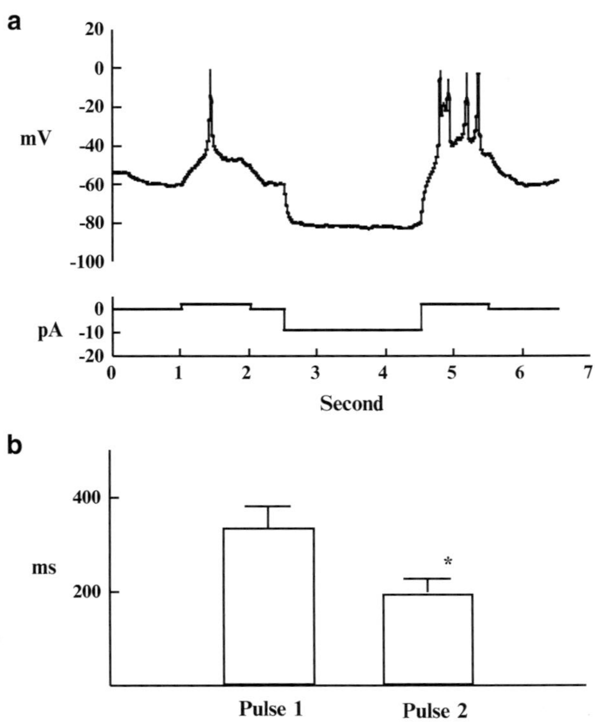

Fig. 10.1 Removal of inactivation of T-type Ca^{2+} channels by a hyperpolarizing current injection increases the frequency of action potentials in β-cells. (**a**) A double-pulse protocol of current injection and corresponding action potentials. (**b**) Comparison of latencies of onset of action potentials between test pulse one and test pulse two

also exhibited a dose-dependent inhibition in glucose-induced insulin release, as measured by radioimmunoassay.

Typical electrical activity of pancreatic β-cells is a series of glucose-induced action potential bursts, which are synchronized with the oscillation of $[Ca^{2+}]_i$ (Cook et al. 1991; Chiavaroli et al. 1991; Gilon and Henquin 1992; Bergsten et al. 1994). Similarly, the oscillation of insulin release is also synchronized with the changes in $[Ca^{2+}]_i$ (Gilon and Henquin 1995). Under non-stimulatory conditions (3.3 mM glucose), rat pancreatic islet cells and a β-cell line also showed spontaneous spikes of $[Ca^{2+}]_i$ in a stochastic and low-frequency manner (author's unpublished observation). Similar phenomena were seen in pituitary cells and neuroblastoma cells, in which the spontaneous $[Ca^{2+}]_i$ spikes were correlated with hormone secretion (Chiavaroli et al. 1992; Gilon et al. 1993). These $[Ca^{2+}]_i$ spikes (and probable electrical spikes) may result from activation of low threshold T-type Ca^{2+} channels. Indeed, activation of T-type Ca^{2+} channels directly triggers synaptic neurotransmitter release at rest (Weiss et al. 2012; Weiss and Zamponi 2013). This newly identified $Ca_V3.2$/syntaxin-1A signaling pathway may also be involved in the mechanism of T-type Ca^{2+} channel-mediated regulation of basal insulin release.

10.4 T-Type Ca^{2+} Currents Modify Basal $[Ca^{2+}]_i$ of Pancreatic β-Cells

Like most mammalian cells, pancreatic β-cells maintain a low level of $[Ca^{2+}]_i$ under non-stimulatory conditions. The values of basal $[Ca^{2+}]_i$ in these cells range from 60 to 80 nM (Longo et al. 1991; Gilon and Henquin 1992; Misler et al. 1992b; Bergsten et al. 1994). In contrast, the average value of $[Ca^{2+}]_i$ that stimulates glucose-induced insulin release ranges from less than 200 nM to 600 nM (Gilon et al. 1993, Gilon and Henquin 1995; Chow et al. 1995). These values are significantly lower than those responsible for neurotransmitter release (10 μM) (Schneggenburger and Neher 2000).

However, it is important to note that in the above studies, the average $[Ca^{2+}]_i$ in β-cells does not take into account the local distribution of Ca^{2+} and thus does not represent the true value of $[Ca^{2+}]_i$ that initiates insulin release. In fact, by activating voltage-gated Ca^{2+} channels, high glucose induces a rise of $[Ca^{2+}]_i$ to a concentration of 8–10 μM directly beneath the cytoplasmic membrane (Soria et al. 2010).

Steady-state $[Ca^{2+}]_i$ in rat pancreatic islet cells, which are chronically treated with a high concentration of glucose (11.1 mM), increases more than twofold, an effect that is significantly inhibited by either mibefradil (1 μM) (Keyser et al. 2014) or siRNA against the α_1G and α_1H subunits of the T-type Ca^{2+} channel (Keyser et al. 2014). Similarly, chronic high glucose (33.3 mM) treatment causes a threefold increase of basal $[Ca^{2+}]_i$ in INS-1 cells (Taylor et al. 2005). The co-incubation of the cells with the T-type Ca^{2+} channel antagonist, NNC 55-0396 (4 μM), largely prevents this effect. Considering the incomplete inhibition of T-type Ca^{2+} channels by these channel blockers at the dosages utilized, the contribution from T-type Ca^{2+} current to the increase in basal $[Ca^{2+}]_i$ should be in the range of 100–200 nM.

10.5 T-Type Ca^{2+} Channels Sustain a Small Current Near the Resting Potential of β-Cells

At low voltages, T-type Ca^{2+} channels can sustain a small current. This phenomenon is known as "window current" (Cohen et al. 1988; Tsien et al. 1998; Crunelli et al. 2005) and is carried by a small population of channels that are not completely inactivated. Biophysically speaking, this current is the result of a voltage overlap between the activation curve and the steady-state inactivation curve at certain membrane potentials. Window current allows the T-type Ca^{2+} channels to regulate Ca^{2+} homeostasis under non-stimulatory conditions (Bean and McDonough 1998). Indeed, T-type Ca^{2+} channels are responsible for the sustained increases in $[Ca^{2+}]_i$ induced by angiotensin II (Buisson et al. 1992), endothelin (Furukawa et al. 1992), and platelet-derived growth factor (Wang et al. 1993). The most direct evidence of T-type Ca^{2+} channel-mediated Ca^{2+} window current is from a study conducted with HEK293 cells expressing the T-type isoform α_1G, which exhibits a window current that peaks at −48 mV (Chemin et al. 2000). The T-type Ca^{2+} channel window current can be demonstrated as a sustained test pulse-elicited current (Monteil et al. 2000). For instance, a two-second sustained Ca^{2+} current of approximately 10 pA was recorded at −55 mV in a rat islet cell (Keyser et al. 2014).

Under resting conditions (~3 mM glucose), human β-cells hold their membrane potentials at approximately −60 mV (Falke et al. 1989; Misler et al. 1992a), which is similar to the value measured in rodent β-cells (Meissner and Schmelz 1974; Meissner and Presissler 1980; Falke et al. 1989; Gilon and Henquin 1992). The resting membrane potential of individual cells can deviate from this value, depending on the activities of K^+ and other channels (Ullrich et al. 1996). At membrane potentials near −60 mV, no other voltage-gated Ca^{2+} current, with the exception of T-type Ca^{2+} channel window current, allows Ca^{2+} inflow. This feature of the T-type Ca^{2+} channel may become more significant when channel expression is upregulated, because upregulated window current will elevate basal $[Ca^{2+}]_i$ in pancreatic β-cells.

10.6 Changes in Basal $[Ca^{2+}]_i$ Regulate Trafficking of L-Type Ca^{2+} Channels On/Off Cytoplasmic Membrane

If changes in basal $[Ca^{2+}]_i$ are far below the level required for stimulating insulin release by glucose, does it have any physiological role in glucose-stimulated insulin secretion?

Yes, the levels of resting $[Ca^{2+}]_i$ determine how β-cells respond to glucose stimulus. During high glucose stimulation, the level of $[Ca^{2+}]_i$ is negatively associated with the level of basal $[Ca^{2+}]_i$ prior to the stimulus (Huang et al. 2004). When cells are perfused with 11.1 mM glucose, a continual increase in $[Ca^{2+}]_i$ ensues as glucose is metabolized. The Ca^{2+} influx resulting from high glucose perfusion was calculated as the peak value of $[Ca^{2+}]_i$ subtracted by the value of basal $[Ca^{2+}]_i$ measured under the pre-stimulatory conditions (3.3 mM glucose). As demonstrated by linear regression, the higher the pre-stimulus basal $[Ca^{2+}]_i$, the smaller the Ca^{2+} influx induced by glucose. Hence there exists a direct inverse relationship between basal $[Ca^{2+}]_i$ and glucose-stimulated Ca^{2+} influx (Fig. 10.2).

The reason underlying the reduction in Ca^{2+} influx in the presence of high basal levels of $[Ca^{2+}]_i$ is due to changes in the availability of L-type Ca^{2+} channels. By employing a combination of whole-cell patch clamp recordings with intracellular perfusion techniques, the effect of $[Ca^{2+}]_i$ on the size of the L-type Ca^{2+} current can be determined (Huang et al. 2004). The replacement of a low $[Ca^{2+}]$ (0 nM) solution by a "high" $[Ca^{2+}]$ (284 nM) through intracellular perfusion resulted in a reduction in the L-type Ca^{2+} current. By comparison, the replacement of a "high" $[Ca^{2+}]$ solution by a low $[Ca^{2+}]$ solution resulted in an increase in the L-type Ca^{2+} current (Fig. 10.3a, b). This effect was attenuated by drugs that interfere with the cytoskeleton, indicating that a channel trafficking process was involved. Furthermore, immunofluorescent labeling of the α_1D subunit of L-type Ca^{2+} channels revealed by deconvolution microscopy an increase in the cytoplasmic distribution of the channels under a high $[Ca^{2+}]_i$ condition (Huang et al. 2004).

Therefore, elevations in the basal $[Ca^{2+}]_i$ of β-cells induce an internalization of L-type Ca^{2+} channels from the cytoplasmic membrane and reduce the amount of Ca^{2+} influx in response to membrane depolarization. This channel internalization mechanism may explain, at least in part, the adverse effect of glucose toxicity on β-cell function.

10.7 Biophysical Kinetics of L- and T-Type Ca^{2+} Channels Are Altered by Changes in Basal $[Ca^{2+}]_i$

In addition to its impact on the trafficking of L-type Ca^{2+} channels, "high" $[Ca^{2+}]_i$ (284 nM) also changes the kinetics of both the L-type and the T-type Ca^{2+} channels (Huang et al. 2004). As shown in Fig. 10.3c, d, "high" $[Ca^{2+}]_i$ mediates a shift toward negative voltages in both the activation of inactivation curves of the L-type Ca^{2+} current.

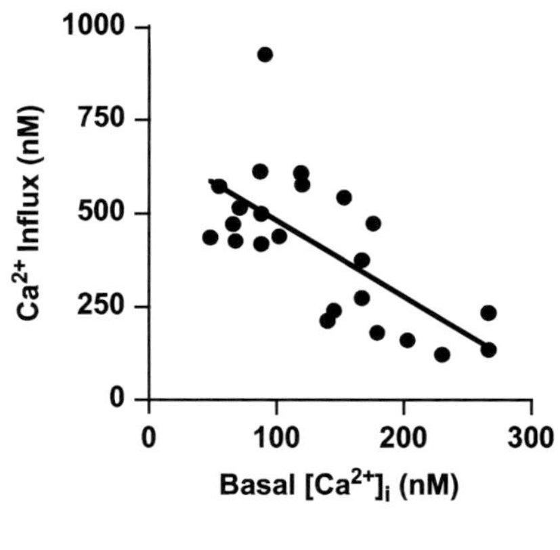

Fig. 10.2 Inverse relationship between glucose-stimulated Ca^{2+} influx and pre-stimulation basal $[Ca^{2+}]_i$. The Ca^{2+} influx is obtained by subtracting the basal $[Ca^{2+}]_i$ measured before high glucose perfusion from the peak $[Ca^{2+}]_i$ after the perfusion. The *line* was constructed by fitting the data by linear regression

However, the activation curve shift seen in Fig. 10.3c is inconclusive since it is presumably masked by a change in the amplitude of the T-type Ca^{2+} channel current. The "high" $[Ca^{2+}]_i$ potentiates the T-type Ca^{2+} current, as indicated by the increase in current at −20 mV and lower. The increase in T-type Ca^{2+} current in the presence of "high" $[Ca^{2+}]_i$ is further confirmed by measuring the slow deactivating tail current at −80 mV after stepping up to −30 mV for 10 ms (Fig. 10.3e). The voltage-dependent activation of the T-type Ca^{2+} channel current was shifted ~11 mV to more negative potentials by a perfusion with medium containing 284 nM Ca^{2+} (Fig. 10.3f).

In summary, an increase in β-cell basal $[Ca^{2+}]_i$ not only internalizes the L-type Ca^{2+} channels from the plasma membrane, but also potentiates the amplitude of the T-type Ca^{2+} channel currents. The enhanced T-type Ca^{2+} currents will, in turn, cause further rising of basal $[Ca^{2+}]_i$. Hence, the T-type Ca^{2+} channel and basal $[Ca^{2+}]_i$ together form a positive feedback loop that can be initiated by an increase in plasma glucose concentration.

The net effect of the upregulation of the T-type Ca^{2+} channel on glucose-stimulated insulin release is difficult to evaluate. On the one hand, it facilitates the general electrical activity of β-cells, thereby promoting insulin release. On the other hand, it causes an elevation in basal $[Ca^{2+}]_i$, which internalizes and inactivates L-type Ca^{2+} channels, thereby inhibiting insulin release. Nevertheless, Fig. 10.2 shows that the inhibitory effect is more dominant after chronic high glucose exposure.

10.8 Changes in Basal $[Ca^{2+}]_i$ Regulate Basal Insulin Release

In response to glucose stimulus, human pancreatic β-cells release insulin via a two-phase process (Porte 1991; Daniel et al. 1999). The early phase of insulin release represents exocytosis of secretory granules in the immediately releasable pool, while the second phase represents exocytosis of the secretory granules that are docked to the cytoplasmic membrane but are not ready for exocytosis (Straub and Sharp 2002). There is also a difference in the signaling pathways that regulate the two phases of insulin release. The early phase is signaled by a K_{ATP} channel-dependent mechanism and the second phase is signaled by a K_{ATP} channel-independent mechanism, which acts in synergy with the K_{ATP} channel-dependent mechanism (Straub and Sharp 2002). Clinical observations show that the major defect of glucose-stimulated insulin release in type II diabetes mellitus is the absence or deficiency of the early phase (Seltzer et al. 1967; Straub and Sharp 2002; Trombetta et al. 2013).

Theoretically, the increase in T-type Ca^{2+} channel activity in β-cells has a double impact on the early phase of glucose-stimulated insulin release. The first effect results in a reduction in L-type Ca^{2+} channels, which conduct Ca^{2+} and trigger exocytosis of insulin granules. This is due to progressive internalization of L-type Ca^{2+} channels associated with elevated basal $[Ca^{2+}]_i$ as described above. As a result, the β-cells that overexpress T-type Ca^{2+} channels show deficient $[Ca^{2+}]_i$-

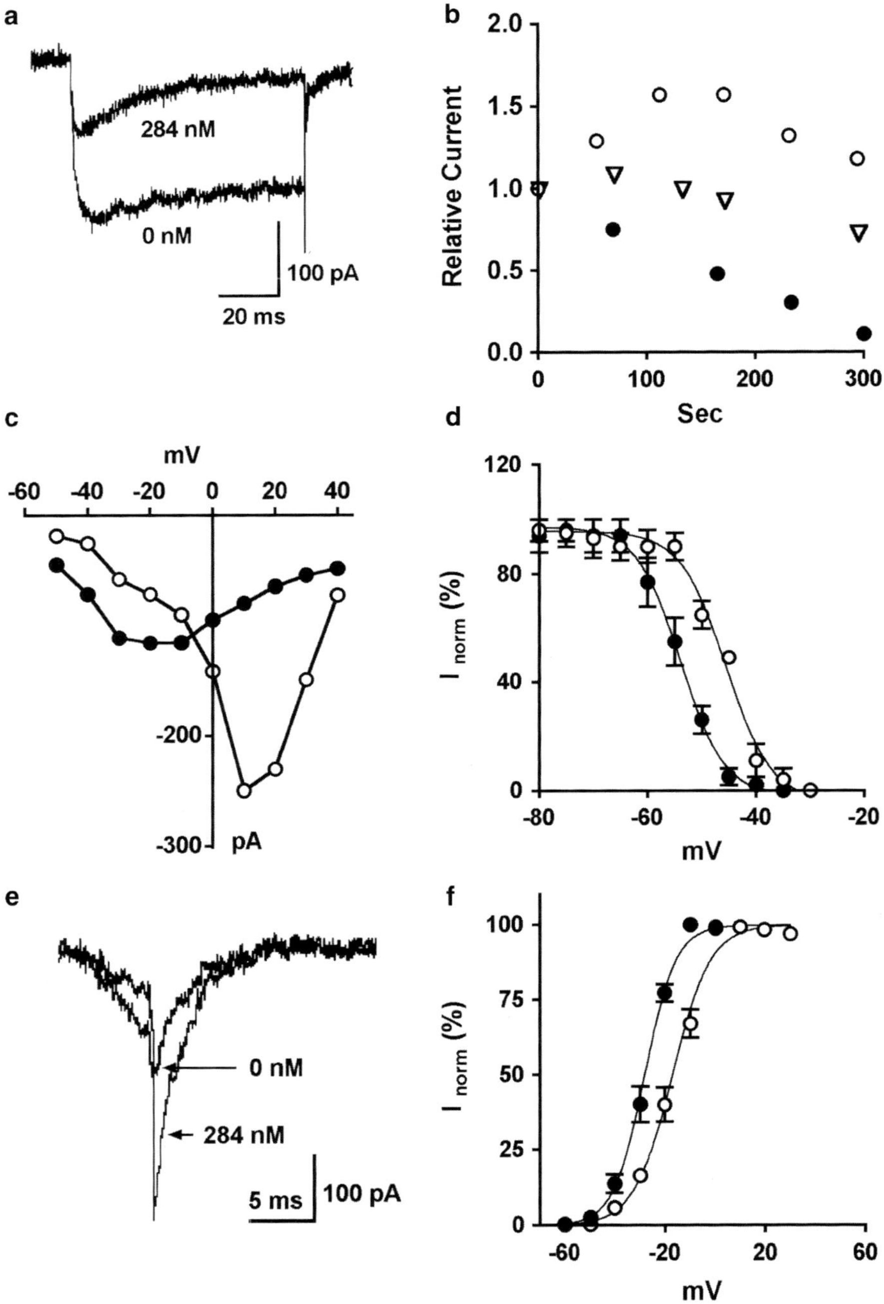

Fig. 10.3 Reversible regulation of L-type Ca^{2+} current amplitude by $[Ca^{2+}]_i$. (**a**) L-type currents recorded before and after an intracellular perfusion of the solution contained 284 nM free Ca^{2+} inhibits the L-type Ca^{2+} currents. (**b**) *Open circles* represent a change in $[Ca^{2+}]$ from 284 to 0 nM, *triangles* represent a change from 0 to 132 nM, and *filled circles* represent a change from 0 to 284 nM. (**c**) and (**d**) The I/V relationship and inactivation obtained after changing $[Ca^{2+}]$ from 0 (*open circles*) to 284 nM (*filled circles*). (**e**) Slow tail currents measured at −80 mV under the

mediated responses to glucose stimulus (Fig. 10.2) as well as reduced stimulus-mediated insulin release. This is similar to the defect in insulin release seen in glucose toxicity. The second impact is a reduction in insulin granules in the immediate releasable pool. As elaborated above, pancreatic β-cells exhibit spontaneous, large $[Ca^{2+}]_i$ spikes in the resting state. These $[Ca^{2+}]_i$ spikes contribute to the promotion of basal insulin release. Increasing the activity of low threshold T-type Ca^{2+} channels may increase the frequency of the spontaneous $[Ca^{2+}]_i$ spikes, resulting in premature release of insulin from the immediately releasable granule pool in the absence of glucose stimulation. Therefore, upregulation of T-type Ca^{2+} channels in β-cells may result in both diminished Ca^{2+} influx through L-type Ca^{2+} channels and premature insulin release, effects that would account for the type 2 diabetic defect in glucose-stimulated insulin release. Since the upregulation of T-type Ca^{2+} channels in β-cells can be stimulated by chronic high glucose exposure, the relationship between glucose stimulus, T-type Ca^{2+} channel activity, and glucose-stimulated insulin release in β-cells can be explained as a positive feedback loop as shown in Fig. 10.4. This positive feedback mechanism may also be involved in β-cell glucose toxicity. Indeed, T-type Ca^{2+} channel antagonists profoundly reduce basal insulin release in cultured islet cells (Keyser et al. 2014) and in β-cells of leptin receptor knockout (db^-/db^-) diabetic mice (Lu et al. 2014).

10.9 The T-Type Ca^{2+} Channel Upregulation Hypothesis of Insulin Resistance

A number of epidemiological studies have demonstrated that insulin resistance exists in the prediabetic state (Weir and Leahy 1994; Dabelea and Hamman 2004; Reaven 2005). During the prediabetic period, pancreatic β-cells produce increasingly more insulin in order to compensate for the development of insulin resistance (Pfeifer et al. 1981).

The production of insulin is inversely proportional to insulin sensitivity (Kahn et al. 1993). The reverse may also be true; insulin sensitivity decreases as the level of circulating insulin rises. In the prediabetic condition, glucose levels fluctuate. The fluctuation of high plasma glucose concentration induced by other concurrent regulatory factors [e.g., inflammatory cytokines (Wang et al. 1999), hypoxia condition (Caravelli et al. 2007), nitric oxide stimulation (Zeng et al. 2005), hormones (Cohen et al. 1988; Ritchie 1993; Michels et al. 2006; Rossier et al. 2003), and genetic predisposition (Lu et al. 2014)] causes increased pancreatic β-cell T-type Ca^{2+} channel expression, which in turn leads to an elevation in basal $[Ca^{2+}]_i$ and an increase in basal insulin release (hyperinsulinemia). Chronic hyperinsulinemia may contribute to the initiation and progression of insulin resistance.

Fig. 10.3 (continued) 0 and 284 nM $[Ca^{2+}]_i$ conditions. (**f**) A negative-shift of voltage-dependent activation of the T-type Ca^{2+} tail currents under 0 (*open circles*) and 284 nM (*filled circles*) $[Ca^{2+}]_i$ conditions

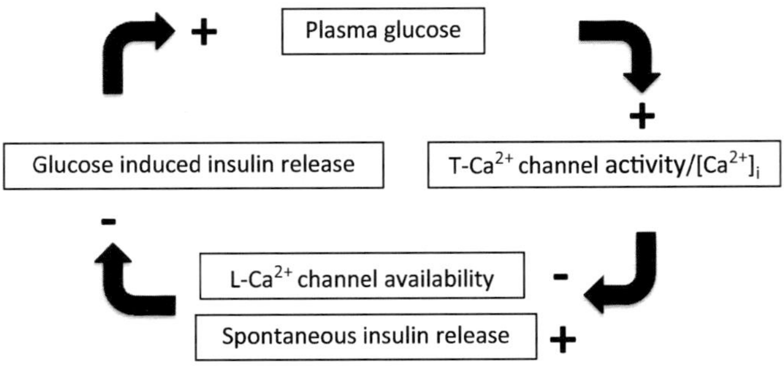

Fig. 10.4 Positive feedback of T-type Ca^{2+} channel activity and plasma glucose concentration. High plasma glucose increases T-type Ca^{2+} channel activity and basal $[Ca^{2+}]_i$; elevated $[Ca^{2+}]_i$ reduces L-type Ca^{2+} channel availability and promotes spontaneous insulin release, resulting in diminished glucose-stimulated insulin release; low insulin release causes high plasma glucose

Conclusions and Further Research

Since the first characterization of T-type channel currents by Armstrong and Matteson (Armstrong and Matteson 1985), the unique biophysical kinetics of these channels have caused scientists to speculate on the most probable function of the channels, which were facilitation of pacemaker potentials of neurons and cardiomyocytes. However, new data reveals that T-type Ca^{2+} channels may also play an important role in regulating $[Ca^{2+}]_i$ homeostasis of a wide range of cells, including excitable and non-excitable tissues. The role of T-type Ca^{2+} channels in regulating basal $[Ca^{2+}]_i$ and basal insulin release in pancreatic β-cells indicates that these channels may also affect the development of type 2 diabetes mellitus and could serve as a new target for antidiabetic therapy.

It has been hypothesized for several years that reducing basal insulin release alleviates insulin resistance (Turner et al. 1979). Exogenous insulin treatment has successfully reversed insulin resistance in humans with early onset diabetes mellitus (Scarlett et al. 1982). A trial comparing early insulin therapy to K_{ATP} channel inhibition in type 2 diabetic patients showed a significant reduction in A_{1C}, as well as improved endogenous insulin secretion (Alvarsson et al. 2003). The treatment of prediabetic patients with T-type Ca^{2+} channels antagonists may correct early disturbances in insulin secretion, thereby preventing or delaying the onset of type 2 diabetes mellitus. Although no human study has been done, initial light has been shed from rodent experiments. Long-term treatment with the T-type Ca^{2+} channel blocker, mibefradil, both prevented and reversed the development of hyperinsulinemia in an insulin-resistant and hypertensive rat model (Verma et al. 1997) and reversed hyperinsulinemia while normalizing glycemic status in diabetic db^-/db^- mice (Lu et al. 2014). Considering that most current hypoglycemic

(continued)

agents do not reverse the advance of type 2 diabetes mellitus, the search for drugs to prevent the onset of the disease is warranted. T-type Ca^{2+} channel blockers are promising candidates belonging to this category and deserve further investigation at the basic and translational level.

References

Alvarsson M, Sundkvist G, Lager I, Henricsson M, Berntorp K, Fernqvist-Forbes E, Steen L, Westermark G, Westmark P, Örn T, Grill V (2003) Beneficial effects of insulin versus sulphonylurea on insulin secretion and metabolic control in recently diagnosed type 2 diabetic patients. Diabetes Care 26:2231–2237

Armstrong CM, Matteson DR (1985) Two distinct populations of calcium channels in a clonal line of pituitary cells. Science 227:65–67

Ashcroft FM, Kelly RP, Smith PA (1990) Two types of Ca^{2+} channels in rat pancreatic β-cells. Pflugers Arch Eur J Physiol 415:504–506

Bean BP, McDonough SI (1998) Two for T. Neuron 20:825–828

Bergsten P, Grapengiesser E, Gylfe E, Tengholm A, Hellman B (1994) Synchronous oscillation of cytoplasmic Ca^{2+} and insulin release in glucose-stimulated pancreatic islets. J Biol Chem 269:8749–8753

Bhattacharjee A, Whitehurst RM Jr, Zhang M, Wang L, Li M (1997) T-type calcium channels facilitate insulin secretion by enhancing general excitability in the insulin-secreting β-cell line, INS-1. Endocrinology 138:3735–3740

Buisson B, Bottari SP, de Gasparo M, Gallo-Payet N, Payet MD (1992) The angiotensin AT2 receptor modulates T-type calcium current in nondifferentiated NG108-15 cells. FEBS Lett 309:161–164

Caravelli V, Marcantoni A, Comunanza V, de Luca A, Díaz J, Rorges R, Carbone E (2007) Chronic hypoxia up-regulates α_{1H} T-type channels and low-threshold catecholamine secretion in rat chromaffin cells. J Physiol 584:149–156

Chemin J, Monteil A, Briquaire C, Richard S, Perez-Reyes E, Nargeot J, Lory P (2000) Overexpression of T-type calcium channels in HEK-293 cells increases intracellular calcium without affecting cellular proliferation. FEBS Lett 478:166–172

Chiavaroli C, Vacher P, Vecsey A, Mons N, Letari O, Pralong W, Lagnaux Y, Whelan R, Schlegel W (1991) Simultaneous monitoring of cytosolic free calcium and exocytosis at the single cell level. J Neuroendocrinol 3:253–260

Chiavaroli C, Cooper DMF, Boyajian CL, Murray-Whelan R, Demaurex N, Spiegel AM, Schlegel W (1992) Spontaneous intracellular calcium oscillations and Gsα subunit expression are inversely correlated with secretory granule content in pituitary cells. J Neuroendocrinol 4:473–481

Chow RH, Lund P-E, Löser S, Panten U, Gylfe E (1995) Coincidence of early glucose-induced depolarization with lowering of cytoplasmic Ca^{2+} in mouse pancreatic β-cells. J Physiol 485:607–17

Cohen CJ, McCarthy RT, Barrrett PQ, Rasmussen HR (1988) Ca channels in adrenal glomerulosa cells: K+ and angiotensin II increase T-type Ca channel current. Proc Natl Acad Sci USA 85:2412–2416

Cook DL, Satin LS, Hopkins WF (1991) Pancreatic B cells are bursting, but how? Trends Neurosci 14:411–414

Cribbs LL, Lee JH, Yang J, Satin J, Zhang Y, Daud A, Barclay J, Williamson MP, Fox M, Rees M, Perez-Reyes E (1998) Cloning and characterization of alpha1H from human heart, a member of the T-type Ca^{2+} channel gene family. Circ Res 83:103–109

Crunelli V, Tóth TI, Cope DW, Blethyn K, Hughes SW (2005) The 'window' T-type calcium current in brain dynamics of different behavioral states. J Physiol 562:121–129

Dabelea D, Hamman RF (2004) Epidemiology of type 2 diabetes mellitus. In: LeRith D, Tayler SI, Olefsky JM (eds) Type 2 diabetes mellitus: a fundamental and clinical test. Lippincott Williams & Wilkins, Philadelphia, pp 785–796

Daniel S, Noda M, Straub SG, Sharp GW (1999) Identification of the docked granule pool responsible for the first phase of glucose-stimulated insulin secretion. Diabetes 48:1686–1690

Falke LC, Gillis KD, Pressel DM, Misler S (1989) Perforated patch recording' allows long-term monitoring of metabolite-induced electrical activity and voltage-dependent Ca^{2+} currents in pancreatic islet B cells. FEBS Lett 251:167–72

Furukawa T, Ito H, Nitta J, Tsujino M, Adachi S, Hiroe M, Marumo F, Sawanobori T, Hiraoka M (1992) Endothelin-1 enhances calcium entry through T-type calcium channels in cultured neonatal rat ventricular myocytes. Circ Res 71:1242–1253

Gilon P, Henquin J-C (1992) Influence of membrane potential changes on cytoplasmic Ca^{2+} concentration in an electrically excitable cell, the insulin-secreting pancreatic B-cell. J Biol Chem 267:20713–20720

Gilon P, Henquin J-C (1995) Distinct effects of glucose on the synchronous oscillations of insulin release and cytoplasmic Ca concentration measured simultaneously in single mouse islets. Endocrinology 136:5725–5730

Gilon P, Shepherd RM, Henquin J-C (1993) Oscillations of secretion driven by oscillation of cytoplasmic Ca^{2+} as evidenced in single pancreatic islets. J Biol Chem 268:22265–22268

Hopkins WF, Satin LS, Cook DL (1991) Inactivation kinetics and pharmacology distinguish two calcium currents in mouse pancreatic β-cells. J Membr Biol 119:229–239

Huang L, Bhattacharjee A, Taylor JT, Zhang M, Keyser BM, Marrero L, Li M (2004) $[Ca^{2+}]_i$ regulated $Ca_V1.3$ translocation in insulin secreting cells. Am J Physiol 286:c213–c221

Iwashima Y, Pugh W, Depaoli AM, Takeda J, Seino S, Bell GI, Polonsky KS (1993) Expression of calcium channel mRNA in rat pancreatic islets and downregulation after glucose infusion. Diabetes 42:948–955

Kahn SE, Prigeon RL, McCulloch DK, Boyko EJ, Bergman RN, Schwartz MW, Neifing JL, Ward WK, Beard JC, Palmer JP (1993) Quantification of the relationship between insulin sensitivity and beta-cell function in human subjects. Evidence for a hyperbolic function. Diabetes 42:1663–1672

Keyser BM, Taylor JT, Choi S-K, Lu Y, Bhattacharjee A, Huang L, Pottle J, Matrougui K, Xu Z, Li M (2014) Role of T-type Ca^{2+} channels in basal $[Ca^{2+}]_i$ regulation and basal insulin secretion in rat islet cells. Curr Trend Endocrinol 7:35-44

Lee JH, Daud AN, Cribbs LL, Lacerda AE, Pereverzev A, Klöckner U, Schneider T, Perez-Reyes E (1999) Cloning and expression of a novel member of the low voltage-activated T-type calcium channel family. J Neurosci 19:1912–1921

Li M, Zhang M, Huang LP, Zhou J, Zhuang H, Taylor JT, Keyser BM, Whitehurst RM Jr (2005) T-type Ca^{2+} channels are involved in high glucose-induced rat neonatal cardiomyocyte proliferation. Pediatr Res 57:550–556

Longo EA, Tornheim K, Deeney JT, Varnum BA, Tilostson D, Prentki M, Corkey BE (1991) Oscillation in cytosolic free Ca^{2+}, oxygen consumption, and insulin secretion in glucose-stimulated rat pancreatic islets. J Biol Chem 266:9314–9319

Lu Y, Long M, Zhou S, Xu Z, Hu F, Li M (2014) Mibefradil reduces blood glucose concentration in db/db mice. Clinics 69:61–67

Meissner HP, Presissler M (1980) Ionic mechanisms of the glucose-induced membrane potential changes in B-cells. Horm Metab Res Suppl 10:91–99

Meissner HP, Schmelz H (1974) Membrane potential of beta-cells in pancreatic islets. Pflügers Arch 351:195–206

Michels G, Er F, Eicks M, Herzig S, Hoppe UC (2006) Long-term and immediate effect of testosterone on single T-type calcium channel in neonatal rat cardiomyocytes. Endocrinology 147:5160–5169

Misler S, Barnett BW, Gillis KD, Pressel DM (1992a) Electrophysiology of stimulus-secretion coupling in human β-cells. Diabetes 41:1221–1228

Misler S, Barnett DW, Pressel DM, Gillis KD, Scharp DW, Falke LC (1992b) Stimulus-secretion coupling in β-cells of transplantable human islets of Langerhans. Diabetes 41:662–670

Monteil A, Chemin J, Bourinet E, Mennessier G, Perez-Reyes E, Lory P, Nargeot J (2000) Molecular and functional properties of the human α_1G subunit that form T-type calcium channels. J Biol Chem 275:6090–6100

Perez-Reyes E, Cribbs LL, Daud A, Lacerda AE, Barclay J, Williamson MP, Fox M, Rees M, Lee JH (1998) Molecular characterization of a neuronal low-voltage-activated T-type calcium channel. Nature 391:896–900

Pfeifer MA, Hater JB, Porte D Jr (1981) Insulin secretion in diabetes mellitus. Am J Med 70:579–588

Porte D Jr (1991) β-cells in type II diabetes mellitus. Diabetes 40:166–180

Reaven GM (2005) The insulin resistance syndrome: definition and dietary approaches to treatment. Annu Rev Nutr 25:391–406

Ritchie AK (1993) Estrogen increases low voltage-activated calcium current density in GH3 anterior pituitary cells. Endocrinology 132:1621–1629

Roe MW, Worley JF III, Tokuyama Y, Phillipson LH, Sturis J, Tang J, Dukes ID, Bell GI, Polonsky KS (1996) NIDDM is associated with loss of pancreatic β-cell L-type Ca^{2+} channel activity. Am J Physiol Endocrinol Metab 270:E133–E140

Rossier MF, Lesouhaitier O, Perrier E, Bockhorn L, Chiappe A, Lalevée N (2003) Aldosterone regulation of T-type calcium channels. J Steroid Biochem Mol Biol 85:383–388

Sala S, Matteson DR (1990) Single-channel recording of two types of Ca^{2+} channels in rat pancreatic β-cells. Biophys J 58:567–571

Scarlett JA, Griffin GJ, Olefsky JM, Kolterman OG (1982) Insulin treatment reverses the insulin resistance of type II diabetes mellitus. Diabetes Care 5:353–363

Schneggenburger R, Neher E (2000) Intracellular calcium dependence of transmitter release rates at a fast central synapse. Nature 406:889–893

Seltzer HS, Allen EW, Herron AL Jr, Brennan MT (1967) Insulin secretion in response to glycemic stimulus: relation of delayed initial release to carbohydrate intolerance in mild diabetes mellitus. J Clin Invest 46:323–335

Soria B, Tudurí E, González A, Hmadcha A, Martin F, Nadal A, Quesada I (2010) Pancreatic islet cells: a model for calcium-dependent peptide release. HFSP J 4:52–60

Straub SG, Sharp GWG (2002) Glucose-stimulated signaling pathways in biphasic insulin secretion. Diabetes Metab Res Rev 18:451–63

Taylor JT, Huang L, Keyser BM, Zhuang H, Clarkson CW, Li M (2005) Role of high-voltage-activated calcium channels in glucose-regulated β-cell calcium homeostasis and insulin release. Am J Physiol Endocrinol Metab 289:E900–E908

Trombetta M, Boselli L, Cretti A, Cali A, Vettor M, Caruso B, Dorizzi R, Avogaro A, Muggeo M, Bonora E, Bonadonna RC (2013) Type 2 diabetes mellitus: a disease of the governance of the glucose-insulin system: an experimental metabolic control analysis study. Nutr Metab Cardiovasc Dis 23:23–30

Tsien RW, Clozel J-P, Nargeot J (1998) Low-voltage-activated T-type Ca^{2+} channels. Adis International Ltd, Chester, England, pp 1–394

Turner RC, Holman RR, Matthews D, Hockaday TDR, Peto J (1979) Insulin deficiency and insulin resistance interaction in diabetes: estimation of their relative contribution by feedback analysis from basal plasma insulin and glucose concentrations. Metabology 28:1086–1096

Ullrich S, Abel K-B, Lehr S, Greger R (1996) Effects of glucose, forskolin and tolbutamide on membrane potential and insulin secretion in the insulin secreting cell line INS-1. Pflügers Arch 432:630–636

Verma S, Bhanot S, Hicke A, McNeil JH (1997) Chronic T-type Ca^{2+} channel blockade with mibefradil in hyperinsulinemic, insulin-resistant and hypertensive rats. Cardiovasc Res 34:121–128

Wang Z, Estacion M, Mordan LJ (1993) Ca^{2+} influx via T-type channels modulates PDGF-induced replication of mouse fibroblasts. Am J Physiol 265:C1239–C1246

Wang L, Bhattacharjee A, Zuo Z, Hu F, Honkanen RE, Berggren P-O, Li M (1999) A low voltage-activated Ca^{2+} current mediates cytokine-induced pancreatic β-cell death. Endocrinology 140:1200–1204

Weir GC, Leahy JL (1994) Pathogenesis of noninsulin dependent (type II) diabetes mellitus. In: Kahn GR, Weir GC (eds) Joslin's diabetes mellitus. Lea & Febiger/Waverly, Philadelphia, pp 240–64

Weiss N, Zamponi GW (2013) Control of low-threshold exocytosis by T-type calcium channels. Biochim Biophys Acta 1828:1579–1586

Weiss N, Hameed S, Fernández-Fernández JM, Fablet K, Karmazinova M, Poillot C, Proft J, Chen L, Bidaud I, Monteil A, Huc-Brandt S, Lacinova L, Lory P, Zamponi GW, Michel De Waard M (2012) A Cav3.2/syntaxin-1A signaling complex controls T-type channel activity and low-threshold exocytosis. J Biol Chem 287:2810–2818

Yang SN, Berggren P-O (2005) β-Cell CaV channel regulation in physiology and pathophysiology. Am J Physiol Endocrinol Metab 288:E16–E28

Zeng X, Keyser B, Li M, Sikka SC (2005) T-type (α1G) low voltage-activated calcium channel interactions with nitric oxide-cyclic guanosine monophosphate pathway and regulation of calcium homeostasis in human cavernosal cells. J Sex Med 2:620–633

Zhang M, Zhuang H, Bhattacharjee A, Li M (2000) High glucose elevated T-type calcium channel expression and basal $[Ca^{2+}]_i$ in rat islet beta cells. Biophys J 78:69A

Zhuang H, Bhattacharjee A, Hu F, Zhang M, Gosmami T, Wang L, Wu S, Berggren P-O, Li M (2000) Cloning of a T-type Ca^{2+} channel isoform in insulin-secreting cells. Diabetes 49:59–64

Index

S.W. Schaffer, M. Li (eds.), *T-type Calcium Channels in Basic and Clinical Science*,
DOI 10.1007/978-3-7091-1413-1

Printed by Printforce, the Netherlands